AF862977

Fascinating Life Sciences

This interdisciplinary series brings together the most essential and captivating topics in the life sciences. They range from the plant sciences to zoology, from the microbiome to macrobiome, and from basic biology to biotechnology. The series not only highlights fascinating research; it also discusses major challenges associated with the life sciences and related disciplines and outlines future research directions. Individual volumes provide in-depth information, are richly illustrated with photographs, illustrations, and maps, and feature suggestions for further reading or glossaries where appropriate.

Interested researchers in all areas of the life sciences, as well as biology enthusiasts, will find the series' interdisciplinary focus and highly readable volumes especially appealing.

More information about this series at https://link.springer.com/bookseries/15408

Tim R. New

Insect Diversity, Declines and Conservation in Australia

Springer

Tim R. New
Department of Ecology, Environment and Evolution
La Trobe University
Bundoora, VIC, Australia

ISSN 2509-6745 ISSN 2509-6753 (electronic)
Fascinating Life Sciences
ISBN 978-3-030-90133-2 ISBN 978-3-030-90134-9 (eBook)
https://doi.org/10.1007/978-3-030-90134-9

© The Editor(s) (if applicable) and The Author(s), under exclusive license to Springer Nature Switzerland AG 2022
This work is subject to copyright. All rights are solely and exclusively licensed by the Publisher, whether the whole or part of the material is concerned, specifically the rights of translation, reprinting, reuse of illustrations, recitation, broadcasting, reproduction on microfilms or in any other physical way, and transmission or information storage and retrieval, electronic adaptation, computer software, or by similar or dissimilar methodology now known or hereafter developed.
The use of general descriptive names, registered names, trademarks, service marks, etc. in this publication does not imply, even in the absence of a specific statement, that such names are exempt from the relevant protective laws and regulations and therefore free for general use.
The publisher, the authors, and the editors are safe to assume that the advice and information in this book are believed to be true and accurate at the date of publication. Neither the publisher nor the authors or the editors give a warranty, expressed or implied, with respect to the material contained herein or for any errors or omissions that may have been made. The publisher remains neutral with regard to jurisdictional claims in published maps and institutional affiliations.

Cover Image: Lord Howe Island Stick Insect or tree lobster (*Dryococelus australis*). Copyright: Rohan Cleave

This Springer imprint is published by the registered company Springer Nature Switzerland AG.
The registered company address is: Gewerbestrasse 11, 6330 Cham, Switzerland

Preface

Interest in insect conservation is founded in concerns over declines and loss of insect species and wider insect diversity, and the roles of human activities in accelerating those disappearances. Recent studies demonstrate or imply that such declines are both too numerous and severe to ignore, and too important for the natural world for us to dismiss or condone. However, and despite abundant evidence of individual species loss, and the unambiguous anthropogenic changes to most natural vegetation types, freshwater environments and other ecosystems throughout the world, with associated changes to their inhabitants, the real extent of taxonomic and functional diversity decline of the less conspicuous animals and plants remains highly speculative. Reflecting their massive taxonomic and ecological variety, insects are major candidates for helping us to clarify the extent of those losses, understand the mechanisms underlying them and appreciate the needs for conservation of all terrestrial and inland water environments.

Despite increasing concerns and advocacy for insect conservation, justification for this to become a global priority remains difficult to communicate and promote, and practical consideration of insect wellbeing still has very low priority on many conservation agendas. Simply, the vital ecological roles of insects in sustaining ecosystems and processes on which humanity (and much of the natural world) depends need wide and powerful emphasis, and to be heeded seriously. Approaches towards remedying this are reasonably well defined and have been discussed extensively, recently by Samways (2020), whose final chapter (on 'Future-proofing the world of insects') is a thought-provoking essay that should become mandatory reading for those whose decisions and activities may either further erode insect life or help to secure it. Many of the steps needed to help save insects are not the exclusive province of 'big science', but can be pursued by individual people and relatively simple means (Kawahara et al. 2021), especially in urban and near-urban environments in which insect losses can be severe, and those measures can have rapid impacts through improving conditions for insect life.

This book provides background to appraising the diversity, vulnerability and needs of insects for more effective conservation in Australia, through focus on the detection and evaluation of insect declines and how to counter them. In part it is also a consolidation of the principles discussed in an earlier book (New 1984), written at a time when the topic of insect conservation was still relatively embryonic in the country and, as for that earlier account, case histories and examples from amongst the better-studied faunas in other parts of the world are vital in setting and advancing the principles and practical conservation management developing progressively for Australian insects. Much of the current practice for insect conservation in Australia has arisen directly from concerns voiced towards the end of the twentieth century, and many publications from that period are still highly relevant in framing later advances as both scientific information and awareness of needs have increased. Subsequent maturation and developments were discussed by New and Yen (2012), with more recent advances noted elsewhere in this book. It is intended for use by students, naturalists, environmental managers and others who may lack specific training in entomology and who may in consequence not appreciate fully the idiosyncrasies, richness and complexity of our largely endemic insect fauna and its vulnerability to environmental changes – as well as the roles of insects in guiding wider conservation endeavour. I have thus included some very basic information as well as more challenging recent developments. The first chapter provides a broad perspective for this treatment, and the problems that arise from attempting to appraise the needs of a diverse and speciose fauna that is still quite poorly documented when compared with insects of parts of the northern temperate region (New and Samways 2014), but for which threats are also varied and severe. Those problems include how to interpret observations of purported changes (Chap. 2) and describe diversity without causing too much confusion (Chap. 3) whilst appreciating that Australia's insect fauna is unique and that the extent of local diversity and the distributions of many species are unknown in other than very general terms. Improving that situation (Chap. 4) acknowledges that simply knowing 'what is there' and 'what has already been lost, and why' are deceptively simple central themes that are almost impossible to address fully. Later chapters turn to practical conservation, from assessing the main causes of declines (Chap. 5), helping to determine the real needs of insect conservation (Chap. 6) and the levels and scope of management for individual species and wider communities in countering threats and reducing future losses to promoting the interest and worth of insect conservation throughout the Australian community (Chaps. 7–10). I have drawn on information available to me up to May 2021. The requirement for each chapter to be read independently, and accordingly referenced individually, means that some overlap is inevitable as key themes and examples are visited in different contexts: page cross-references are used throughout the text to facilitate broader use.

Many non-entomologist readers working in environmental management and seeking guidance on insect conservation may be led to contrast the paucity of basic information on insects in Australia with the far greater (but, still for many taxa manifestly inadequate) data and interest levels available for conservation policy and practice for many other Australian biota, notably amongst endemic vertebrates and vascular plants that may gain far higher public sympathy. Australia is amongst the rather few countries with a megadiverse insect fauna and which is also sufficiently 'rich' that their conservation can be taken responsibly. Some of the problems of appraising insect diversity in Australia (discussed also by New 2011) remain formidable challenges, both for species enumeration and preparation of reasonably definitive inventories, and plots of distribution. These constraints necessarily underpin attempts to evaluate insect faunas, and are amongst the major impediments to doing so and to appraising faunal changes as environments change, either naturally or from human impacts on the wealth of natural environments that allowed Raven and Yeates (2007) to proclaim that 'Biodiversity defines the way that Australia looks and the way that it functions'. Native insects are key components of that vision.

Melbourne, Australia Tim R. New

References

New TR (1984) Insect conservation – an Australian perspective. W Junk, The Hague

New TR (2011) 'In considerable variety': introducing the diversity of Australia's insects. Springer, Dordrecht

New TR, Samways MJ (2014) Insect conservation in the southern temperate zones: an overview. Austral Entomol 53:26–31

New TR, Yen AL (2012) Insect conservation in Australia. In: New TR (ed) Insect conservation: past, present and prospects. Springer, Dordrecht. pp 193–212

Raven PH, Yeates DK (2007) Australian biodiversity: threats for the present, opportunities for the future. Aust J Entomol 46:177–187

Samways MJ (2020) Insect conservation. A global synthesis. CABI, Wallingford

Acknowledgements

I am very grateful to the following for allowing me to reproduce material to which they hold copyright: Council of Europe Publishing, Strasbourg; CSIRO Publishing, Melbourne; Elsevier; IUCN (the World Conservation Union); John Wiley and Sons; and the Department of Conservation, New Zealand. Every effort has been made to obtain these permissions, and the publishers would appreciate notice of any inadvertent omissions or corrections that should be included in any future editions or imprints of this book. Figures used have been redrawn to standardise lettering, and the individual legends indicate any other modifications. The cover picture of a Lord Howe Island Stick Insect was taken by Rohan Cleave, and I very much appreciate his permission for this to be used here (credit: Rohan Cleave, Zoos Victoria).

At Springer, Dr Lars Koerner, as Editor, has enthusiastically supported this enterprise, and I greatly appreciate his ideas, advice and practical help as the book developed. My Project Manager, Mr. Bibhuti Sharma, guided later production of the book. The careful efforts by Mr. Arunkumar Kathiravan in editing and setting the book are greatly appreciated.

Contents

Chapter 1
Introduction: The Reality of Insect Declines

1.1 Reasons for Concern

The central importance of conserving insects and their relatives is stlll not accepted fully. Many people consider them—if at all—as nuisances or pests largely inimicable to their own interests and wellbeing, and otherwise of little importance. Losses or declines of insects and other invertebrates may then cause little concern, and even be welcomed as numbers of nocturnal visitors, biting flies and others diminish. This unfortunate perception—based on only a tiny proportion of insect species—is a major barrier to gaining support for insect conservation, and links strongly with lack of awareness of the enormous variety of insects and understanding of what they contribute to the natural world, as well as their vulnerability to environmental changes wrought through human activities. Thus, 'the general public's disregard for insects is a significant obstacle to insect biodiversity conservation' (Duffus et al. 2021), and overcoming this is a key component of future conservation strategy. A major national overview (Yen and Butcher 1997) and a large workshop gathering of invertebrate specialists in Australia (Cassis 1999) both identified the widespread lack of understanding of values of invertebrates amongst decision-makers and the general public as a fundamental barrier to pursuing conservation.

Exceptions occur. Butterflies are more widely accepted than most other insects as in some way 'worthy', some large spectacular dragonflies are equally conspicuous and also attractive to many people, and parallel acceptance occurs sporadically in some other groups such as beetles, cicadas and stick insects. A few, indeed, have become important 'conservation icons' in various parts of the world, from either or both of aesthetic and cultural appeal. Declines, particularly of butterflies as the most generally popular insect group, have been the most important stimulus to wider concerns for insect conservation, emanating especially from parts of the northern temperate regions where the fauna has been studied for a century and more, and the accumulated specimen records and collector documentation has clearly demonstrated changes in distributions and abundance of many species. Many of those

© The Author(s), under exclusive license to Springer Nature Switzerland AG 2022
T. R. New, *Insect Diversity, Declines and Conservation in Australia*, Fascinating Life Sciences, https://doi.org/10.1007/978-3-030-90134-9_1

trends can be associated with changes to their natural environments as human activities and influences expand and diversify and well-publicised restricted butterfly habitats (including, especially in Europe, 'traditional' collecting localities for many rare species) have succumbed to development. Decades of concern have engendered models for emulation elsewhere as similar anthropogenic impacts occur.

However, insects have far greater significance than simply whether we like them or not. The wider needs to conserve them and what they do in the natural world reflect both their enormous diversity and the range of their ecological roles. Simply, the insects are among the greatest manifestations of animal life forms that has ever existed, occur in most terrestrial and freshwater environments, and are predominant participants in most food webs and ecological processes. The numbers of different insect species are unknown and some of the suggested figures, their limitations, and how they may be refined progressively, are discussed later in this chapter. Clearly, they vastly outnumber most other animal groups for which estimates are available and this predominance is evident also in their abundance and biomass. Indeed, they have for long been considered by far the most diverse of all animal groups, but in a recent reappraisal Larsen et al. (2017) suggested that insects are really only one of three hyperdiverse animal groups, the others being mites (Acarina) and nematode worms (Nematoda), both far more poorly studied than insects. They continue to diversify, but many of the distinctive and consistently recognisable key groups of insects have been around for several hundred million years—itself a mark of outstanding evolutionary success—and the global insect fauna has been continually 'edited' by diversification and losses over these long periods. Nevertheless, some modern insects clearly reflect very ancient lineages and appear to have changed little from their presumed ancestors in the fossil record; some, indeed, are referred to as 'living fossils' for that very reason.

In addition to their intrinsic interest and the basic ethical belief that no life forms should be exterminated by human cupidity, losses of insects have severe practical consequences for ecosystems and ecological processes. Samways (2020) described the insects as 'the major shareholders of biodiversity', and their central roles in food webs and wider community interactions are increasingly apparent. Some years ago they, together with other invertebrates, were described as 'webmasters in ecological systems' (Coleman and Hendrix 2000), and the term 'ecosystem engineers' has also been applied repeatedly. This term aptly fits some exceptionally abundant insects, perhaps most notably the social ants and termites whose activities can markedly change soil characteristics and nutrient pathways. They exemplify well that insect conservation must encompass the superabundant species and those that are at present widespread and common as well as the more familiar focus on rare or threatened species that have already declined. Perhaps this ecological importance has become most widely evident through well-publicised and widespread declines of pollinating insects (notably bees and some flies), leading to crop failures and major concerns over future food supplies for people. Such 'ecosystem services' have universal relevance—disruptions to natural pollination systems, and changes to trophic interactions including herbivory, predation, more complex trophic interactions, decomposition and nutrient cycling—many of which are specific and collectively sustain the ecosystems of the world—can have wide-ranging 'cascade effects'. Whilst references to insects simply as 'bird food' and the like seem disparaging, they relate to their vital central roles in food webs—and a number of recent insect declines have been discussed in conjunction with declines of insectivorous vertebrates. Many such roles are essentially unseen. They go largely unnoticed until they slow or cease, when those roles may come forcibly to wide attention, for example with much of the concerns for pollinators emanating from the agricultural and horticultural industry sectors as this vital service fails. Other insects are also significant contributors to human activities. Numerous small wasps are 'parasitoids', using other insects as hosts in which to develop, and many insect groups are predators on other insects. Both these interactions come to notice in the context of pest management, with those insects categorised as 'natural enemies' that can contribute to suppression of pest insects on crops and commodities as biological control agents. Much modern agriculture seeks to increase this application by using native insects as part of 'conservation biological control' associated with lessening of pesticide use and more intensive agricultural practices.

The sheer numbers of insect individuals and species, and the multitude of interactions in which they participate—many of them not investigated in detail, and their variety changing continually—renders consequences of their loss highly uncertain. For example, the specific pollinators of many endemic Australian plants are unknown, so that unheralded losses of poorly known bees, beetles, flies, moths or others might lead to plant extinctions. As plant-pollinator mutualisms are lost, changes to plant communities are likely to affect numerous other insect herbivores, together with their predators and parasitoids, in turn affecting nutrient recycling patterns as well as the structures of many food webs. Many insects are indeed ecological specialists with highly specific diets and other environmental requirements. As well as being restricted to particular regions, climatic regimes and environments (loosely, their 'habitats'), their very specific food needs necessitate equally specific interactions with other life forms and with each other. Insect species, in general, are not interchangeable. Changes to the local environment, including climate change, can alter the 'balance' and intricacies reached over long evolutionary time, perhaps with catastrophic effects within the communities. E.O. Wilson's (1987) well-known aphorism, that if insects and their relatives disappeared the world as we know it would no longer exist, recognises the importance of insect life and of taking urgent measures to sustain it and, as far as possible, to prevent losses. In the European Strategy for invertebrate conservation, Haslett (2008) listed a number of reasons why their conservation is important (Table 1.1). These are based on knowledge of the world's best documented regional invertebrate fauna—of which insects are a major component—and transfer easily to Australia as a compelling summary of their roles.

Increased scarcity of some butterflies and beetles has been publicised through collector interests or concerns, with increasing efforts needed to obtain specimens and the 'black market' for some rare species continuing to proclaim their availability—at inflated prices that reflect the legal and practical difficulties of obtaining

Table 1.1 Summary of the major reasons for conserving invertebrates, as listed by Haslett (2008)

Invertebrates:
1 are key components of, and perform essential functions in just about every ecosystem
2 contribute much more to biodiversity than any other group of organisms on our planet
3 dominate in nearly every conceivable habitat, but are the same time extremely vulnerable
4 are the basic food resource, directly or indirectly, for other organisms higher in the food web
5 provide a valuable and still underestimated resource for commercial use in medicine, technology, food and other services
6 have aesthetic value that remains often unappreciated and contributes significantly to the beauty and enjoyment of nature

representatives of species recognised as endangered or otherwise threatened. In that sense of demand, those insects become commodities, a role shared with others that are collected from the wild for some practical benefit or use. In parts of inland Australia, increased tourism has led to increased interest in 'bush tucker', the traditional wild-collected food of the first Australians. This includes the 'witchetty grubs', large larvae of wood-feeding moths and beetles, collected individually and which might become susceptible to overharvest in places.

Insects deserve conservation both for what they are and what they do. Losses of insect diversity beget losses of ecological functions, so that maintaining insect variety and abundance is a major facet of sustaining ecosystem structures and functions in any given geographical area. The unique nature of many Australian environments gives them global significance. Their losses would assuredly erode massive unseen endemic biodiversity, among which insects are a predominant component. Appraising the scale of the conservation need objectively is a continuing challenge—but enhancing public and political acceptance of the values of this enterprise is a basic foundation for progress. Difficult though it may be to communicate the needs to conserve this multitudinous, poorly unknown, difficult to diagnose and largely unappreciated low-profile array, and with very little fundamental information to hand on the needs of most taxa, sensible precaution dictates that this aim is an important aspect of the future of Earth's ecosystems and its human inhabitants. To quote from a recent major review (Dasgupta 2021) in referring to increasing extinction rates of biodiversity, 'Such declines are undermining Nature's productivity, resilience and adaptability, and are in turn fuelling extreme risk and uncertainty for our economies and well-being'.

1.2 Assessing Insect Declines

Declines of insects are not only a recent phenomenon, but the extent to which they have been inferred and reported over the last decade or so has apparently surprised many people. Historical extinctions or local disappearances of individual species, together with more widespread declines, are well-documented among the best studied taxa and areas—perhaps most notably amongst butterflies in parts of western Europe (Habel et al. 2016) for which species extinctions in particular countries were evident in the nineteenth century—and there is little reason to doubt that parallels are widespread in other taxa and in most parts of the world. Many older naturalists in Australia can nostalgically recall insect-related 'scenes of their childhood' when, perhaps most notably in the decades before public transport and car ownership reached current levels, urban and near-urban environments supported far greater insect life than is nowadays apparent. That anecdotal history can only rarely be validated rigorously, but is a salutary signal of declines in several 'popular' insect groups—examples include putative declines of cicadas from roadside trees, Christmas beetles (*Anoplognathus* spp., Scarabaeidae), Emperor gum moths (*Opodiphthera* spp., Saturniidae), and others, mostly large and conspicuous taxa associated with native vegetation. Parallel trends have been reported widely elsewhere, and declines of insect numbers on car windscreens, and at street and store lights in urban and near-urban areas have prompted strong warnings over consequences of such losses. In those cases, the losses are not confined to individual taxa but span total insect biomass and abundance across many taxa and ecological roles. Decreasing incidence of insect pollinators of key field and orchard crops, and diminished insect food supply for insectivorous birds and bats, for examples, gain frequent mention as symptoms of the breakdown of 'ecological services' and erosion of natural food webs, and losses of insect variety from areas subsumed by intensive agriculture, expanding urbanisation and other environmental despoliation—all potentially exacerbated by climate change—have extended well beyond intensively settled areas to the more remote parts of 'protected areas' in the anthropogenic landscape (Janzen and Hallwachs 2019).

A rather different form of change in Australia, again paralleled elsewhere, has been suggested as declines of some regular insect migrants, perhaps most notably the vast southward spring migrations of Painted lady butterflies (*Vanessa kershawi*, Nymphalidae) to southern Victoria and of the later seasonal migrations of the Caper white butterfly (*Anapheis java teutonia*, Pieridae), both discussed by Braby et al. (2021). Both these butterflies reflect phenomena sufficient to have elicited public comments in the past, with published migration records extending over many decades. They are paralleled by the recent apparent severe decline in numbers of Bogong moths (*Agrotis infusa*, Noctuidae, p. 22) migrating to aestivate in the higher elevation areas of the south-east mainland. For all, a combination of landscape changes and climate influences have been suggested as contributory. Disturbances that lead to changes in such notable behaviour may be far more frequent, but detected only amongst such highly visible incidences across the landscape. The Brown awl butterfly (*Badamia exclamationis*, Hesperiidae) formerly underwent spectacular mass migrations associated with summer breeding in Queensland. However, the system has declined since the mid-1990s—Valentine (2004) used the term 'demise', and noted that the clearing of vegetation in central Queensland, notably of the vast tracts of Yellowwood (*Terminalia oblongata*, Combretaceae) trees for cropping and pasture was the most obvious factor affecting larval food supply and, in all likelihood, the abundance and migrations of *Badamia*. Many possible similar occurrences in other orders have not been recorded—but this does not necessarily mean they have not occurred, so again emphasising the immense gaps in understanding insect biology in relation to landscape structure and changes. Indeed, the widespread changes in seasonal migration patterns reported for insects in Costa Rica (Janzen and Hallwachs 2016) are likely to reflect much more general incidence across the world that can be affected substantially as habitats and climate change and influence the resources needed. Loss of taxonomic and ecological variety of insects as landscapes and water bodies are transformed reflects not only the changed spatial living conditions but also the isolation of those environments as they become progressively less hospitable to and reachable by many previous residents. In short, the changes to vegetation systems and landscapes wrought by human activities have rendered many insect species (and other invertebrates) increasingly vulnerable to loss.

Many insect species have very restricted and specific environmental needs, such as specific foodplants, prey or hosts, as well as physical and climatic conditions, which contribute to allocation of many such species as ecological specialists with limited geographical and ecological distributions. Changes to those conditions may increase their vulnerability, through altered ecological interactions, tolerances and opportunities. At extremes, diverse botanical assemblages may be replaced by crop monocultures, in which any increase in plant diversity is countered by chemical or cultural suppression, entire forests or woodlands may be cleared for housing or industrial developments, and regulation or drainage of wetlands cause enormous changes to the conditions needed by numerous specialised native insects. Should the environment become untenable or unstable, such an insect has three major options: it can (1) change to adapt to the new conditions, and persist; (2) move away to discover and colonise other areas in which its requirements persist or have developed and where it can form sustainable populations; or (3) become extinct as no longer able to survive in the changed environment and lacking capacity to adapt or disperse. Most insect species that are known to have become extinct are indeed ecological specialists with very restricted ecological needs and/or limited dispersal ability, or for which alternative breeding areas are not available or are inaccessible because of distance or inhospitable landscapes. However, it is believed widely that most insect extinctions go unnoticed, and that the relatively few confirmed by recent studies are only a tiny proportion of those that have occurred. Thus, Fonseca (2009) referred to the 'silent mass extinction' of herbivorous insects in Earth's (then) 34 biodiversity hotspots.

Even in the best-known local insect faunas, detailed histories of loss or decline are largely restricted to a few 'popular' orders or other larger groups (notably those with historical collector/hobbyist interest and documentation, such as butterflies and some larger moths, beetles, dragonflies, and some grasshoppers and aculeate Hymenoptera), providing a strong taxonomic bias to existing knowledge (Leandro et al. 2017). Most other insect orders are far more inadequately known, have relatively few devotees, and their losses are only rarely remarked, unless fortuitously reported. Many species are naturally scarce, rarely observed, and their demise unlikely to be noticed or admitted as 'real'—except in cases in which key habitats on which specialised species depend are destroyed. This scenario extends widely to other invertebrates, with losses in many phyla and orders both likely and unobserved. Very broadly, for most of these taxa, available information is simply inadequate to detect losses or population declines, and distributions and ecological relationships of most species are insufficiently known—indeed many of the species themselves are still to be discovered and described. It is believed widely that the dearth of insects and other invertebrates on lists of 'threatened species' does not imply that losses are not occurring, but rather that they are not being detected and their reality investigated. As Rix et al. (2017) put it 'poorly known invertebrates are disappearing "silently", sometimes without having ever been described'. Again, despite considerable ambiguities over interpreting many studies, 'Overall, Europe has seen a significant and ongoing loss of general insect species richness' (Habel et al. 2019), manifested not only as reduced richness but also declines in abundance of many species and shifting composition of species assemblages.

Much evidence for insect declines is predominantly from taxon-focused surveys, many of individual species and others of a few key higher taxa amenable to long-term or otherwise repeated surveys and monitoring. However, declines of individual taxa may not necessarily mirror a more widespread and general trend affecting a far greater variety of insects (Hallmann et al. 2017). Superimposed on this species-based scenario is one in which overall insect diversity, of both abundance and richness of both common and rare species encompassing a full range from generalist to specialist taxa, is also declining. Analysis of Malaise trap catches of airborne insects (trapped abut 1 m from the ground) in Germany implied decline of more than 75% in flying insect biomass occurring over the 27 years of catches examined. The traps were sited in 63 protected nature areas, although the number of multiple samples from individual sites was low. Hallmann et al. (2017) suggested that the severity of the declines demonstrated that those declines extended beyond those insects conventionally considered to be vulnerable, and were distributed widely across taxa. Using only those sites at which traps were present for at least 2 years gave rates of decline similar to those from the entire trap set, and Hallmann et al. considered the declines representative of lowland protected areas in west Germany. Declines were not obviously correlated with habitat type or landscape features: both lack of clear attribution to definable cause(s) and that they occurred in protected areas are major concerns, together with the likely cascade effects that might flow for the region's major ecosystems. Warnings such as that generated from insects in Dutch nature reserves, in which a range of different taxa appeared to have undergone considerable declines (Hallmann et al. 2020), supporting those found for Germany, support the case for the far more widespread losses discussed by Sanchez-Bayo and Wyckhuys (2019, 2021). Collectively, those studies embrace many different sampling approaches, data sets and taxa but, as these authors and others discuss, the actual causes and extents of declines can only rarely be defined, and more detailed monitoring and ecological insights are needed for this. Insect declines are recognised as a major concern both for ecological functioning and integrity and for human health and welfare (van der Sluijs 2019; Cardoso et al. 2020). They demand urgent attention, but decisions are thwarted by highly incomplete knowledge of their reality, extent and causes—and it is impracticable to await full understanding before conservation is actively pursued. Thus, Harvey et al. (2020) proposed a series of immediate 'no-regret' measures to slow or stop declines (p. 70) and urged that 'We must act now'. Janzen (in Janzen and Hallwachs 2019) put it more forcibly '. . . when a house is burning down, I need the fire department to keep it from spreading, not a thermometer to measure the obvious heat and flames', and emphasised again that 'The house is burning. We do not need a thermometer. We need a fire hose'. Many entomologists would agree unhesitatingly, but establishing that acceptance and need within societal decision-making, especially in the tropics, is a herculean task. Solutions now available to help conserve insect populations, such as those suggested by Cardoso et al. (2020), mostly emphasise removing the major known causes of declines and loss—very broadly encompassing preserving and restoring natural habitats, eliminating some agricultural practices (such as use of harmful pesticides), taking measures to counter or avoid effects of harmful invasive species, taking steps to drastically reduce greenhouse gas emissions as the major cause of climate change, and curbing effects of overexploitation of insects. Such attractive and relevant general recommendations can become very complex to pursue.

Severe declines amongst previously abundant species are perhaps of especial concern, although the fate of very scarce species is often far more widely publicised. Declines of many previously abundant insect pollinators, for example, have attracted wide attention and concern, not least because of their roles in assuring human food supply (Potts et al. 2016). Wagner (2020) regarded declines of those species conventionally regarded as 'common' until recently and across numerous species as potentially able to 'cascade' across food webs and lead to degradation of ecosystem services. Insect declines are about much more than ethics and aesthetics, and their ramifications can strongly influence human welfare and needs. Reductions in population size, so that insects become less abundant rather than disappearing completely, are viewed commonly as a step on the road to extinction, and may also link with contractions or fragmentation of geographical range. Perceived declines across whole insect assemblages, both in biomass and abundance, are justifiable causes for alarm, and many of the long-term surveys leading to those claims are from Europe or North America. Analysis of 166 long-term studies, across a total of 1676 sites, showed considerable variations in trends, even between adjacent areas (van Klink et al. 2020). Intriguingly, that survey inferred that terrestrial insect abundance had declined by about 9%/decade, but abundance of freshwater insects had increased by around 11%/decade, possibly reflecting improvements in water quality whilst terrestrial environments continue to be degraded more.

Recent commentaries on insect declines have stimulated considerable attention and debate, ranging from discussions of the assumptions and scientific information used to more impetuous hyperbole. Although founded in genuine concerns for 'biodiversity' and in some cases motivated to increase awareness of concern, inevitable dangers arise from reduced credibility as accounts are queried or invalidated. This confusion can backfire and lead to dismissal of very serious concerns. As Didham et al. (2020) wrote, the unprecedented public attention flowing from such evidence 'might be a double-edged sword if the veracity of alarmist insect decline statements do not stand up to close scrutiny'. Any claims of insect declines merit serious attention. Irrespective of their extent or magnitude, they may signal losses that affect ecosystem dynamics and human welfare, as well as the plight of Earth's biodiversity. The variety of putative declines reported transcends methodologies, sampling intensities and duration, and taxa to imply that serious losses continue. Debates over the magnitude of insect declines have tended to dominate recent reactions to some published estimates of losses, but arguments over that extent do not mask the reality that declines are indeed occurring across many taxa, regimes and ecosystems. Nevertheless, the dilemmas over assessing the severity of insect losses can strongly influence political agendas and public awareness. Some assessments and caveats are noted here.

Claims for very major losses of taxa—the 'Insectageddon' of Thomas et al. (2019)—need very careful appraisal. This situation has been termed the 'Insect Armageddon Narrative', with media claims of dire consequences for Earth's ecology, including claims even that extinction of all insects is faced within a hundred years, sensationalism addressed by Saunders et al. (2020a, b). Importantly, such drastic declines were not explicitly stated in the major papers by Sanchez-Bayo and Wyckhuys (2019) and two other significant contributors (Hallmann et al. 2017; Lister and Garcia 2018). Accompanying media releases from the home institutions of the lead authors extrapolated the published results to imply far greater levels of loss. As Saunders et al. (2020a) emphasised, it is then a small step to that extrapolated on social media channels, with some subsequent uncritical citations based solely on these sources. Later inclusions in peer-reviewed papers afford credibility based simply on earlier uncritical extensions from more authoritative studies.

A major science-based stimulus to that concern claimed that extinction of up to 40% of the world's insects could occur within the next few decades, and that those losses span terrestrial and freshwater ecosystems and the full range from generalist to specialist insect species (Sanchez-Bayo and Wyckhuys 2019). That meta-analysis included review of 73 reports of insect declines, from many parts of the world, and attempted to interpret the major causes ('drivers') of those declines. Publications reviewed were retrieved from an online database (the Web of Science) using three keywords for search, namely [insect*] AND [declin*] AND [survey], from which a total of 653 publications were retrieved, and information complemented by publications referred to in each. However, the subset of 73 selected reports encompassed those that reported quantifiable changes in either abundance or species richness over time.

Later commentators (Komonen et al. 2019) pointed out biases in this selection: 'If you search for declines, you find declines', and imprecise definition of drivers reflected in simply enumerating the examples in which a particular driver was implicated. In short, concerns over the general validity of interpretation by Sanchez-Bayo and Wyckhuys have been expressed by several concerned respondents and, in turn, have stimulated further contributions to interpreting insect declines based on what Didham et al. (2020) assessed as 'less extreme claims of a global crisis, and providing more robust approaches to detail and evaluate population and richness trends'. More recently, Sanchez-Bayo and Wyckhuys (2021) presented additional examples of population changes, to bring their cumulative assessments from 100 long-term studies on insects, mostly from the last 34 decades and with that information from the 10 major insect orders suggesting that an average of 37% of species were declining in numbers.

Needs for more robust interpretations include advocacy for clearer and more temperate language to avoid both ambiguity and misleading impacts on policy and wider debate. Thus, Saunders et al. (2020b) noted that even the word 'decline' may often be inappropriate because its connotation is commonly negative, and suggested four key points to consider if its use is contemplated: (1) specificity of use to the taxon or study referred to rather than implying wider unproven generality; (2) whilst much debate has implied that decline is global, most studies refer to only restricted areas, from locality to nation, and often to only particular biomes or ecosystems: the context should be specified; (3) studies should not have a primary aim to detect decline but, rather, deal with population trends or differences or the impacts of environmental characteristics on population dynamics, using appropriate methods and heeding biases in sampling procedures; and (4) evaluating what the results/studies actually show—many studies across taxa and/or localities reveal a variety of declines, increases or stability across these, so that 'decline' alone may convey a misleading impression.

Nevertheless, many cases of insect declines have been reported (Wagner 2020), and widespread evidence of insect losses based on sound documented evidence are grounds for serious concern. The main caution is not to over-sensationalise an already substantial and urgent reality by extrapolating from ambiguous or distorted evidence, and thereby weaken the message that must be heeded at high levels if conservation is to be pursued properly. In short, the media attention that Saunders (2019) referred to as 'a heady mix of hyperbole, anecdote, and speculation' can create confusion and problems for validation and countering extravagant claims. Didham et al. (2020) succinctly commented that 'greater care is needed when evaluating evidence from population levels and in identifying drivers of those trends'. Others have also queried the basis on which apparently dramatic declines have been postulated. But, in short, the flurry of publications from recent critics and contributors has led to strong inferences that—although insect declines are indeed widespread—assumption of a 'global armageddon' are probably misleading, but perhaps only premature (Cardoso and Leather 2019; Showalter et al. 2019; Simmons et al. 2019), and also emphasise the biases in data interpretation and how these may be remedied or countered. Schowalter et al. urged that careful attention be paid to the biases in the various sampling methods and the care needed in interpreting generally available 'open access' data sets, but their comment that 'establishing insect abundance is very sensitive to sampling frequency and duration' has universal relevance. As examples of planning needed, Montgomery et al. (2020) remarked that there are steps that can indicate putative declines properly, and Didham et al. (2020) considered seven 'challenges' that should be addressed in attempting to do this (Table 1.2). These measures, however wise and urgent, represent enormous endeavour, and other workers (Forister et al. 2019) have urged more immediate action based on what is already known and understood. They point out, correctly, that delays attending

Table 1.2 Seven challenges to drawing robust inferences about insect population declines, identified by Didham et al. (2020)

1. Establishment of the historical baseline
2. Representativeness of site selection
3. Robustness of time series trend estimation
4. Mitigation of detection bias effects
5. Ability to account for potential artefacts
6. Phenological shifts
7. Scale-dependence in extrapolation from sample abundance to population-level inference

greater accumulation of information will almost inevitably cause many further losses of insects.

The seven challenges listed in Table 1.2 effectively expose major biases in some appraisals of insect population declines, and emphasise also that insect populations are highly dynamic and can change dramatically (over several levels of abundance to decrease or increase in numbers) over very few generations, and may be influenced heavily by local environmental conditions and circumstances. Proper assessments that can justify the occurrence of declines depend on standardised sampling over periods sufficiently long to reveal that trend set against a valid baseline of more natural fluctuations and ecological context, with taxa identified reliably and the geographical area specified. Thus, one major bias noted by contributors following the claims by Sanchez-Bayo and Wyckhuys (2019) is that the great majority of studies they reviewed are from Europe or North America, with few (if any) representing the rest of the world—including the entire tropical regions and southern temperate zones. This bias is entirely realistic, in reflecting information accumulated over long historical traditions of natural history interest and insect collecting over up to almost 200 years, as traditions absent elsewhere. Indeed, modern insect conservation has its origins largely over concern for butterflies in Britain and western continental Europe, where declines have largely reflected habitat changes imposed by people, and records of presence and abundance are generally sufficient to track such changes. Thus, for Britain mapping distribution of all species and plotting ranges over time on a relatively small scale across the countries has provided an enormous amount of reliable information on butterflies and demonstrated numerous changes (Asher et al. 2001). Nevertheless, as Sanchez-Bayo and Wyckhuys (2021) noted, their assessment perhaps underestimated the losses that are occurring in those regions with massive environmental changes involving forest losses and intensifying agriculture. However, amidst reported declines of species, some other taxa have increased. Beirne's (1955) overview of fluctuations in abundance of British Lepidoptera, in which he attempted to summarise findings from several thousand published references, included a comment that those collected records 'indicate fluctuations that often range from conditions when few or no specimens are seen in a locality for several years to conditions of great abundance for the same species there', but also that 'statistical data are relatively few'. 'Collector lore' accumulated through those numerous records provide the best available historical background to insect abundance, but by more modern standards is not ‘robust’, however valuable as indicating how changes occur. Long-term declines are often difficult to distinguish from the short-term ‘noise’ of normal population fluctuations. Nevertheless, the long-term standardised surveys needed to confirm changes in insect abundance and diversity robustly are rare; notwithstanding needs for long-term monitoring, those studies are expensive and difficult to sustain and for many parts of the world simply do not exist.

Thus, determining a historical background of equivalent integrity from which to estimate more recent changes is rarely possible elsewhere. ‘In the tropics, where most insects live, circumstantial data exist, but long-term records are too sparse to support a conclusion of a global insect decline’ (Basset and Lamarre 2019). Debates—emotive or more informed—on ‘cataclysmic declines’ of insects can give way progressively to more perceptive practical conservation, based in part on distinguishing the causes of changes in richness, abundance of individual species and distributions. Drivers of all are complex (Wilson and Fox 2020) and interact in many ways. These authors urged the importance of effective monitoring of insect responses to conservation measures and environmental changes, with one benefit being to foster political and societal interest and support. Failures to monitor conservation operations over sufficiently long periods to attribute outcomes reliably are very frequent in insect conservation, and can commonly not distinguish a treatment response from usual numerical fluctuations in the populations.

Determining such a reliable historical baseline as advocated above is only rarely possible. Didham et al. (2020) noted that any perception of that baseline depends on the extent and validity of historical knowledge. Any baseline proposed is thus ‘shifting’ (so that even when a historical point is selected, the position of that point on any trend gradient is unknown) and ‘false’ (where such a point is at an abundance peak, so that subsequent declines may simply be part of a normal correction toward a population mean level). Selection of study sites within the focal species’ range leads to what Didham et al. termed the ‘missing zero effect’, that unoccupied sites that are potentially suitable (for example within a metapopulation structure) will be under-represented in local studies. Rates of decline can be overestimated at sites where abundance is measured, but may not consider movements leading to concurrent increase at other sites, whether unoccupied or previously unoccupied, with the factors influencing the kind and consequences of population fluctuations not clearly understood. The ‘snapshot effect’ reflects the frequent approach to measuring change, of comparison across two time points, ‘historical’ and ‘recent’, as a single paired comparison. Either more time points or more individual sites are needed to truly validate this approach and to assess whether the difference between two points is in any way a realistic reflection of change. Inherent in all the above themes is an assumption that the sampling undertaken is relevant and standardised, and that the detectability of the target insect(s) is valid and consistent, through both suitable methods and applying them properly with awareness of the insects’ ecological and behavioural repertoires. Related biases can flow from inadequate knowledge or understanding. Pitfall trapping, for example, as a method used extensively to sample ground-dwelling insects, and the visual surveys used commonly for butterflies and dragonflies, tend to measure activity rather than abundance, and the outcomes reflect factors such as site topography or weather conditions. Findings may be limited to the actual site sampled, if the variety of such caveats is not considered in comparisons.

A further problem discussed by Didham et al. is that of 'temporal autocorrelation' in the detectability of individuals, whereby abundance may appear to change simply because individuals become more or less detectable as sampling proceeds. Such variations in apparency can occur over various time scales: short term changes in cloud cover, temperature, wind speed, precipitation, or even the time of day (p. 201) may affect activity and, hence, detectability. Ability to consistently detect, capture and/or count individuals can thus differ greatly under different conditions, and commonly also within a restricted seasonal window to find the target, and can change in unexpected ways. For low abundance species, additional care may be needed because the sampling programme itself can reduce abundance through removal of individuals, and cause local depletion. Species with low dispersal ability and low reproductive rates may be especially susceptible: one representative case is pitfall traps leading to local depletion of large-bodied flightless ground beetles (Carabidae), or when their sex ratio changes over a survey period. Movements of some Odonata away from waterside areas where visual surveys are conventionally undertaken may involve females, only, moving away for weeks or months to mature, giving the appearance of a lowered population with a strongly biased sex ratio that then 'suddenly' increases and changes in structure as individuals return.

Other effects have also been advanced to support the claim of decline. Some are discussed later, but two examples mentioned, earlier are (1) lower numbers of moths at lights and (2) lower numbers of insects caught in vehicle radiator grills and squashed on windowscreens. For the first, the increasing number of anthropogenic light sources, especially in urban and periurban areas may cause 'dilution' of catches as moths and other phototactic insects (Owens et al. 2020) are spread across numerous individual lights rather than concentrated at fewer attractive sources so that some doubters have implied that numbers have not actually reduced but individuals present have become more widely dispersed. Lower numbers of insects in car grills and on windows have been attributed simply to designs of more aerodynamic vehicles rather than any actual difference in insect numbers killed (Vogel 2017); evidence for this is highly uncertain, and many commentators hold that declines inferred in both the above contexts are indeed real, and evocative personal accounts such as that by Janzen (Janzen and Hallwachs 2019) justifiably cause alarm. Any such implications of declining insect abundance are of concern and, if real, may have far-reaching impacts. Perhaps only for pollinators have such declines been investigated in detail. Bees are among the most intensively studied insect groups and accounts of declines in bumblebees (*Bombus* spp.) and the fate of domestic honeybees (*Apis mellifera*) convey widespread concerns over human food supply.

More generally, abundance trends for individual species can be detected or inferred only from well-documented insect groups or particularly notable and characteristic species in other orders, for which any historical information facilitating long-term interpretation is invaluable. Many purported declines in recent years lack any long-term baseline. A major exception is the larger Lepidoptera in western Europe (particularly in Britain), for which information is based on decades of systematic sampling using light traps operated by recorders across the countries. From that vast data set, Fox et al. (2019) used more than 11 million records of species occurrences over 1970–2010 to assess changes in frequency of 673 'macromoth' species in Great Britain. Their restrictions included that growth of new biological information and detection methods over that period could influence detectability, so that biases were countered by (1) excluding non-resident species and those subject to taxonomic revision since 1970, and (2) excluding species for which recording methods had changed—notably of most clearwing moths (Sesiidae) because of increased use of pheromone attractants in recording them. The remainder still included about 10 ½ million records for analysis. From these, 260 species had declined significantly, whilst 160 species increased significantly. Declines were not restricted to scarcer species of reported conservation concern. Earlier, Conrad et al. (2006) had noted that many moth species regarded as 'common and widespread' in Britain were undergoing severe population declines. At that time, 71 species were declining at rates sufficient to accord them formal 'endangered' or 'vulnerable' status under the IUCN criteria. Further, none was a notable long-distance migrant, so that it was unlikely that British populations could be 'rescued' by migrants from continental Europe. Collectively, about two thirds of the 337 species examined by Conrad et al. declined over the 35 years covered by their account, and these represented a wide range of biotopes. The catches analysed comprised species for which >500 individuals were captured over the total period, and annual national indices for the 199 sites that were operated for at least 48 weeks/year for 5 years. Fox (2013) commented that those studies, in conjunction with others from continental Europe (Mattila et al. 2006; Essens et al. 2017; Groenendijk and Ellis 2011, the last implying that about 70% of moth species in the Netherlands had declined in abundance, many to the level of being considered threatened) provided 'the first evidence of national scale declines in a species-rich insect taxon' and, whilst individual species continued to be a strong focus, such wider declines may be far more widespread than generally acknowledged. Parallel evidence for bumblebees (Goulson et al. 2008; Kosior et al. 2007; Cameron et al. 2011) based on substantially fewer records has also been advanced.

Both Lepidoptera and bumblebees demonstrate the scale of concern, involving entire local faunas, and the trends revealed from the few long-term studies available are mirrored by numerous short-term recent comparisons. As a key example, changes in the butterflies and diurnal burnet moths (Zygaenidae) of a region of Germany were assessed from records accumulated over 1840–2013, to reveal that the proportion of habitat generalist species increased, whilst many specialist species with specific habitat needs, restricted dispersal ability and rather higher threat levels were extirpated (Habel et al. 2016). Those losses led to recent species assemblages being largely subsets of those present earlier, and this trend was shown also in the higher nitrogen inputs leading to increasingly eutrophic larval hosts with losses of some more specialised habitats such as calcareous grasslands. Kosior et al. (2007) assessed the distributions and threats to all 60 known species of bumblebees and cuckoobees (*Psythirus* spp.), as an important group of pollinators, across 11 countries of western and central Europe from the beginning of the twentieth century, from a combination of published and unpublished records, their own observations, and questionnaires completed by experts from those countries. They also compartmentalised the information to contrast the first and second halves of the century. Altogether, six and 13 taxa respectively became extinct in at least one country over those periods, and four species became extinct in all 11 countries from 1951 to 2000. Collectively, 29 country-level extinctions were detected, with 21 of these in the later period. Only 20% of the species were deemed not to be threatened in some way, with others assessed as threatened in at least one country—and, many, more widely. The two possible reasons given for the greater extinction rates from 1951 were (1) less accurate information from the earlier period and (2) increased level and variety of threats. As for Britain (Williams 1989), changes in agricultural practices were considered a strong influence.

A similar but more restricted survey in North America (Cameron et al. 2011) used historical collection records of bees and comparison with a recent (2007–2009) field survey over 382 sites across the United States. For the eight selected target species, totals of 73,759 museum specimens and 16,788 recent specimens were available. Four of those species had apparently declined, and the others had remained reasonably stable. Range-wide declines in the former group were implied to have occurred over recent decades but other, co-occurring, species remained widespread and common. As for Europe, therefore, declines in some taxa were accompanied by other taxa remaining more stable in both abundance and distribution. However, Cameron et al. postulated that their sampling perhaps did not confirm real losses but, rather, that the species had become sufficiently rare to render them very difficult to detect.

Widespread awareness that such declines may not be unusual among common and ecologically generalised insect species contrasts with earlier opinions that only the more specialised and 'rare' species are likely to be lost. As Wagner (2020) emphasised, this changed perspective raises serious concerns because abundant insect species can have significant roles in food webs across all major biomes and their loss may indeed induce some form of 'ecological breakdown' of functional resource supplies that influence numerous other taxa. Whilst species richness itself can change over historical time, abundances of both specialist and generalist taxa may also alter substantially. Thus, over 160 years of observations and collection records of Lepidoptera in Germany, six trend categories were recognised by Laussmann et al. (2021). Two of these categories ('not rated', for reasons such as recent taxon recognition or uncertain data; 'single observations' over the period) were not analysed further but the other four—with constituent species categorised as specialists, generalists or neither—showed the pattern summarised in Fig. 1.1. High proportions of specialist taxa were either constantly rare or declining, whilst few generalists occurred in either of those categories, with many of them constantly

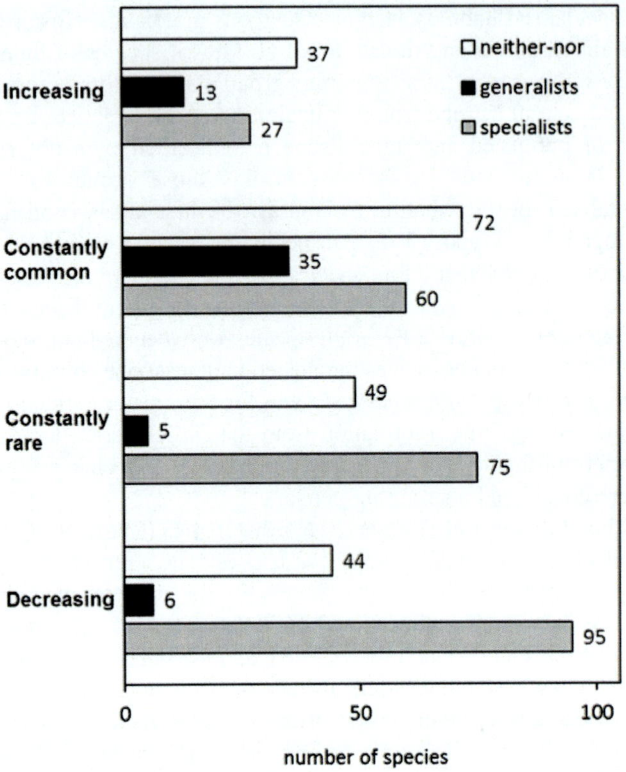

Fig. 1.1 Trends in abundance of butterflies and larger moths (together 'macrolepidoptera', including 537 evaluated species) in the Wuppertal region of Germany over 160 years. Bars shown are numbers of species in each of four major trend categories (after Laussmann et al. 2021)

common. However, the number of specialist species (276) greatly exceeded numbers of generalists (78).

Wilson's (1987) widely cited account of insects being among the 'little things that run the world' is brought forcefully to notice with the documented breakdown of pollination systems through losses of insect pollinators, and implications that parallel erosions across other key interactions are also likely to occur. Declines of previously common species, both in abundance and distribution have become a key theme in insect conservation, augmenting the more familiar concentration on individual rare or threatened species. The above examples reflect the greater attention possible within the best-known insect faunas, and some commentators have indeed queried whether similar losses occur elsewhere, in places from where they have not been reported. Wagner (2020), considering only terrestrial insects, found very few examples beyond the above regions—perhaps largely reflecting lack of documentation. Elsewhere, problems of simply enumerating, evaluating and identifying the insect fauna become formidably difficult, and reliable documentation of systematics, ecology and distribution remains relatively incomplete. For many places, the chances of improving this situation to approach levels taken for granted elsewhere are slim. Those general problems are outlined in the next chapter, within the context of massive and poorly known diversity.

That 'Human activity is responsible for almost all current insect population declines and extinctions' (Cardoso et al. 2020) is a maxim of considerable relevance, even when assessing the extent and rate of those losses is difficult due to inability to recognise and enumerate many of the species involved. For example, Stork (2018) considered that only 20% of the world's (perhaps) 5.5 million insect species are named—and of these only a tiny fraction (1) have been evaluated for conservation need or (2) are known in detail sufficient to attempt that evaluation. Recent accounts and media attention drawing attention to insect extinctions have increased public awareness but in many cases these simply recapitulate and endorse concerns long held by biologists but which have been largely ignored at the levels of formulating conservation policy and priority. As Dunn (2005) noted, the study of insect extinctions has been neglected—to the extent that he reported only 70 documented extinctions, although 'thousands are estimated to have occurred'. That most of those extinctions were among the better-documented faunas of North America and western Europe simply reflected that only there can unambiguous evidence enable their detection. The vast scale of likely losses of insects, whether extinctions or substantial declines, and the wider consequences of these, is still to become sufficiently acknowledged. Lack of direct evidence over most of the rest of the world renders detecting losses largely speculative—but inferred strongly as vegetation is cleared, lakes and rivers polluted, and the needs of burgeoning humanity take precedence over documenting and conserving the natural world. Failure to enumerate insect species fully is by no means confined to Australia and, indeed, few regions of the world can boast even reasonably complete insect inventory. Even deciding which of the four 'megadiverse' holometabolous insect orders (namely Coleoptera, Diptera, Lepidoptera and Hymenoptera) is the most speciose is still debatable, and claims continue to be advanced. Although widely stated that Coleoptera are the most diverse, there is increasing suggestion that this accolade stems largely from beetles having been for long a popular group for collectors, so that they have a strong historical record of interest and numerous species have been described: as hard-bodied, they are also easier to preserve than more 'delicate' insects. However, both Hymenoptera (Forbes et al. 2018) and Diptera (Borkent et al. 2018; Brown et al. 2018) have earned convincing advocacy for their predominance. The richness of Hymenoptera rests largely among the enormous array of mostly tiny parasitoid wasps, of which very numerous species remain to be described and for which levels of host specificity are largely undocumented. Even for Britain, substantial gaps remain in understanding host-parasitoid interactions and conservation needs (Shaw and Hochberg 2001). Appraising, as far as possible, the parasitoid:host ratios from a series of relatively well studied natural interactions in North America, the most conservative estimates found suggested that Hymenoptera outnumber Coleoptera by more than 1.3 times (Forbes et al. 2018). Richness of Diptera is also vast, as demonstrated by recent documentation of the flies found in a four-hectare tropical cloud forest site at Zurqui de Moravia in Costa Rica (Borkent et al. 2018; Brown et al. 2018), yielding an unexpectedly high richness through thorough taxonomic examination by a large team of specialists and that included the usually neglected small and delicate flies that are commonly simply discarded as 'too difficult' to deal with. Across 73 families, 4332 fly species were distinguished, representing more than half the number of named Diptera from Central America, 14% of all named Neotropical flies and about 2.7% of the global named Diptera. Clearly, diversity of Diptera in the tropics is enormous, and links with ecological variety: 'Diptera . . . are remarkably diverse in terms of species, morphology and ecology' (Borkent et al. 2018). The diversity of flies at Zurqui was not considered unusual, and two families were especially rich in species. Cecidomyiidae (gall midges, with 800 species found) and Phoridae (with 407 species) both yielded enormous numbers of specimens and even more species were probably present. They were dominated, respectively, by *Mycetophila* (136 species) and *Megaselia* (240 species) (Brown et al. 2018), both genera among the most 'difficult' of all Diptera to identify to species level. For perspective, only four named species of *Mycetophila* were known previously from Costa Rica.

Many more surveys of flies, 'microhymenoptera' and other insects from a range of tropical sites spanning different ecosystems and biogeographical regions are needed to even suggest more realistic numbers of species present, and improve understanding of insect diversity. At present, 'It is difficult to relate to non-specialists how vast our ignorance truly is' (Borkent et al. 2018). Estimates of overall insect diversity, especially in the tropics, remain very uncertain, but future surveys combined with molecular level interpretations (Hebert et al. 2004) continue to raise estimates—thus, Garcia-Robledo et al. (2020) emphasised that lack of the fundamental data on insect richness also renders detection and evaluation of declines uncertain and largely impracticable as 'vast cryptic diversity', especially among small flies and parasitoid wasps, continues to be exposed. Such immense cryptic diversity extends well beyond tropical faunas. Haplotypes of a million insects sampled from Malaise traps across Canada (Hebert et al. 2016), for example, implied that more than 16,000 species of gall midges alone were present in Canada, and that the global total for that family alone could approach two million species. Clearly, much uncertainty persists over the 'real numbers' of insect species and the ranking of richness across the major holometabolous orders, with greatest uncertainties among the tiny flies and wasps among which 'more is unknown than known' (Garcia-Robledo et al. 2020) and that are in due course likely to raise recognised global totals far above the estimates of 5–10 million insect species discussed by Stork (2018). However, at a practical level, rather than satisfying human need for discovery and precision, Janzen (1994, 1997) emphasised that simply counting species is somewhat irrelevant because we already know that there are a lot, many of them are poorly known, narrowly distributed and perhaps rare, and that their habitats are vulnerable to anthropogenic changes and, in many cases, disappearing rapidly. Reiterated by later workers (including Borkent et al. 2018), 'counting species' is at that level alone necessary basic documentation for scientific inquiry but does not further advance conservation other than in setting firmer baselines against which extinction rates might be assessed.

Dunn considered that two components of insect extinction might be especially important: (1) the extinction of narrow habitat specialists, as a widespread topic (even, paradigm) across insect conservation, and (2) the co-extinction of dependent species with their hosts, a theme far less prominently acknowledged. Loss of the specific ectoparasites of vertebrates, for example, may not be uncommon, but in all possible Australian cases only the vertebrate host species has been recognised formally as extinct (Woinarski et al. 2015). They noted the case of two endemic rats from Christmas Island that became extinct around 1904 together (presumably) with their two endemic ectoparasites, a tick (*Ixodes nitens*) and a flea (*Xenopsylla nesiotes*, Pulicidae). These are acknowledged as extinct but are yet to take their place on any formal list or compilation of Australia's extinct taxa. The flea, specific to *Rattus macleari*, was considered 'the first reported case of the extinction of a host-specific flea' (Kwak 2018). That unenviable status for unheralded ectoparasites was for long accorded to the two lice of the Passenger pigeon (*Ectopistes migratorius*) in North America, and both the lice were listed by IUCN as Extinct (E) following demise of that formerly abundant host. However, both species have now been found on living pigeons (Dunn 2005). Eight Australian fleas were considered of conservation significance as their hosts decline. Two are specific to the critically endangered (CE) Leadbeater's possum (*Gymnobelideus leadbeateri*, Victoria's faunal emblem and a flagship for conservation of old-growth eucalypt forests in central Victoria), and were ranked as CE to reflect the parlous state of their marsupial host (Kwak 2018). One of these, the Wiradjuri flea (*Wurunjerria warnekei*, Pygiopsyllidae), indeed, may already be extinct.

The complexities of conserving host-parasite relationships are commonly overlooked in practical conservation, and the extreme host-specificity of many insect ectoparasites of vertebrates not appreciated fully. Efforts to conserve an endangered mammal or bird through captive breeding may involve sanitation measures such as 'delousing' animals as they are removed from the wild and subsequent treatments to rid them of any further parasites. Conversely, conservation of an ectoparasite may be anticipated by introducing the parasite to captive hosts, with the longer-term prospect of release to the wild from reared stock (Dougherty et al. 2016). Four possible outcomes for such parasite conservation were projected by Dougherty et al., depending on whether either party was secure or extinct. Under ideal conditions both host and parasite are conserved successfully. Alternatively (1) the parasite alone may be lost, perhaps especially on endangered hosts that themselves persist in only low numbers; (2) the parasite survives through being found on other hosts after the original host is lost; or (3) both parties become extinct. All these scenarios have been implied (Rosza and Vas 2014) and both 'co-extinct' and 'co-critically endangered' parasitic lice have been distinguished. The latter included 40 (or 41) lice as critically endangered because their bird or mammal hosts were categorised thus by IUCN. Comments on extinct lice included (1) *Colpocephalum californici* (Menoponidae) which became extinct during the captive breeding and release programme for the Californian condor (*Gymnogyps californianus*), possibly from veterinary delousing treatments; (2) the New Zealand *Rallicola pilgrimi* (Philopteridae), which became extinct when the Little spotted kiwi (*Apteryx owenii*) was translocated to predator-free islands; and (3) two species of *Coloceras* (Philopteridae) regarded as co-extinct with the Norfolk Island pigeon, *Hemiphaga novaeseelandiae*. Deliberate destruction of lice in captive host populations reflects the widespread belief that they increase host vulnerability through sickness or stress, but destruction of parasites before host release may restrict their variety in the reintroduced populations. Nevertheless, conservation management for vertebrates has commonly led to parasite losses. Host-specific parasites may, indeed, become more endangered that their host (Stringer and Linklater 2014), for example when the host is the more widely distributed partner, and local parasite control measures may then increase the extinction risk.

A rather different form of potential coextinction involves a small group of oecophorid moths (p. 111) whose larvae feed on the excreta of parrot nestlings in nests excavated in termite mounds (p. 184) in northern Australia, and which are not known from any other habitats. The three moth species (one of which is thought to be extinct) have been recorded from different parrot hosts, and it is unclear whether the association is mutualistic or commensal. The Antbed parrot moth (*Trisyntopa scatophaga,* Oecophoridae) is known only from nests of the Golden-shouldered parrot (*Psephotus chrysopterygius*), and is wholly dependent on this host which is itself rare, declining and assessed as 'endangered'—not least from illegal capture of parrots for the aviculture industry. The fate of *T. scatophaga* seems linked inextricably with that of the parrot, which nests only in mounds of *Amitermes* and, more rarely, *Nasutitermes triodiae*; its biology was discussed by Edwards et al. (2007). The main host termite, *Amitermes scopulus*, is endemic to parts of Cape York Peninsula, as is the parrot. A related moth, *Trisyntopa neossophila*, is associated with the Hooded parrot (*Psephotus dissimilis*) in the same region (Cooney et al. 2009). *T. scatophaga* has been found in only one small area of Cape York but is likely to be more widespread and probably co-incident with *P. chrysopterygius* across the five areas where it occurs. The main recommendations of a conservation plan proposed for *T. scatophaga* are focused on measures to conserve the parrot host: as a highly novel step, Edwards et al. (2007) pondered whether aviary breeding stocks of both the above parrot hosts might be a valid approach to conserving the moths in conjunction.

Coextinction of phytophagous insects with decline of their specific host plant is exemplified by the recent apparent extinction of an Australian mealybug, *Pseudococcus markharveyi* (Hemiptera: Pseudococcidae), discussed by Moir (2021). Described only in 2013, the mealybug suffered coextinction in 2019–2020. Its sole host plant, the critically endangered and highly localised *Banksia montana*, had been reduced substantially by fires (which kill the shrubs) and the alien root fungus *Phytophthora cinnamomi*. *P. markharveyi* was known from only two montane ridges in Western Australia's Stirling Ranges National Park, from two *Banksia* populations about 7 Km apart. All individual plants known to harbour the mealybug were burnt in a fire from December 2019, and Moir (2021) considered it likely to have become extinct. Predation by native ladybirds

(Coccinellidae) may have contributed to earlier declines. Significantly, as Moir commented, loss of *P. markharveyi* occurred despite substantial conservation interventions, including translocations, over the previous 8 years. Any such direct and obligate association may imply that extinction could occur, with 'co-extinction' of the partners implicit in specific plant-pollinator interactions. Those partners may differ substantially in their longevity and biology—so that a long-lived plant may 'outlive' loss of its pollinator by decades or more until its death, perhaps creating an illusion that 'all is well'. A second context of secondary extinctions, extinction cascades, represent wider extinctions that flow from single extinctions through subsequent interactions within food webs so that the impact is extended to an array of interacting species, is far more complex to detect and analyse (Kehoe et al. 2020). Concerns thereby extend to potential losses of species that are only indirectly associated with the initially lost taxon.

As implied above, very few invertebrates (10) are listed formally as extinct in Australia, and these included only two insects (Woinarski et al. 2019): a Tasmanian caddis fly (the Great Lakes Caddis fly, *Costora iena*) and the Lord Howe Island ground weevil (*Hybomorphus melanosmus*). The latter, in particular, exemplifies a wider concern—that a further nine endemic Lord Howe Island beetles have not been seen for at least a century and are presumed to be extinct. Numerous parallels on the Australian mainland span many insect orders with species that, even through targeted searches by specialists, have not been seen since their initial discovery. Many of these are inferred to be (or to have been) narrowly distributed, ecologically specialised and likely to participate in restricted interactions with other species, so vulnerable to co-extinction. These are also simply those species that have been recognised: many others have not yet been diagnosed or even detected (Chap. 2). Implications for Australian insect losses and conservation are discussed later (Chap. 6).

Fonseca (2009) estimated the numbers of monophagous (that is, 'specialist' through being restricted to a single host plant species) insect herbivores in the world's then 34 biodiversity hotspots identified as the top priorities for conservation interests (Mittermeier et al. 2004), and reflected in the high levels of plant endemics as the foundation of wider biodiversity. Each hotspot supported at least 1500 endemic plant species, and many of them had already undergone considerable loss and changes. Strong correlations were found between numbers of monophagous insect herbivores and the numbers of endemic plants. Fonseca's models revealed that for habitat loss within the hotspots with contraction of host range the estimated numerical range of monophagous insect species committed to extinction from losses of their endemic host plants was an alarming 213,830–547,500 species—figures alarming for both their magnitude and uncertainty, but perhaps not unrealistic. If endemic plant genera were included, so that oligophagous insects (those restricted to a plant host genus) were also included, these numbers increased even further. Other similar attempts to assess numbers of susceptible phytophagous insects in part reflect combining the numbers of endemic plants and the average numbers of insects associated with each, extending from the hypotheses initiated by Erwin (1982). For the various hotspots, Fonseca (2009) used a conservative average of 5.3–10.7

monophagous species/plant, contrasted with the higher numbers of 10 (Gaston and Hudson 1994) and 23–37 (Basset et al. 1996). Every such attempt has suggested that very large numbers of plant-associated specialist insects could become threatened, but that honing any such number to even the nearest hundred thousand species is stlll very uncertain.

Essentially, levels of extinction among insects (and other invertebrates) are simply unknown. Hochkirch (2016) commented that 'estimates vary from 1 to 100 per day', and uncertainties flow directly from lack of basic taxonomic and ecological knowledge, together with ignorance over local species richness and distributions of the various taxa: consequently 'insect extinctions go unnoticed'. For most insect orders and many subordinal taxa the notion of 'species' as consistently recognisable discrete entities is itself highly uncertain, although 'species numbers' are an almost universal currency in insect conservation debates.

Nevertheless, insects 'despite their almost ungraspably large numbers, are at disproportionate risk of extinction' (Berenbaum 2009), and the level of that disproportion remains an open question. Estimating any reasonably accurate rate or extent for insect declines in Australia is difficult, reflecting lack of information, but anecdotal evidence suggests that a number of urban/periurban insects were far more abundant only a few decades ago—declines of 'popular' species such as Emperor gum moths (*Opodiphthera* spp., Saturniidae), cicadas and Christmas beetles (*Anoplognathus* spp., Scarabaeidae) have all been implied by naturalists. For a few other 'icon species' dramatic recent declines have aroused wider concerns, and it is uncertain to what extent these are reflected in wider insect communities or guilds. The Bogong moth (*Agrotis infusa*, Noctuidae) has attracted considerable attention, as one of Australia's most iconic insects, long famed for its extensive migrations (of more than 1000 km) to aestivate in sheltered cave and rocky retreats in the alpine areas of the south east and for its cultural significance to local Aboriginal nations, to whom the vast numbers of resting moths—their over-summering aggregations can reach densities as high as 17,000 moths/m^2 (Common 1954)—were an important food resource (Flood 1980). It is also a key food item for the endangered Mountain pygmy possum, *Burramys parvus*, and low moth numbers have been associated with low survival rates of possum offspring. Concerns arose from a severe 'crash' of moth numbers in spring 2017, from around 4.4 billion moths (one account suggested 8.8 billion!) in alpine areas to very few individuals, in a decrease continuing also in the following year. Suggested causes included lack of rainfall during winter drought and climate change impacts on the lower elevation breeding areas, as well as changes in farming practices such as (1) increased clearing of land for crops and (2) increased insecticide uses. Climate changes may lead to suitable aestivation sites also being only at higher elevations. One of the frequent inducements for interest in the moth has been its attraction to lights, so that large numbers of moths invade venues such as sports grounds and public buildings, including Parliament House in Canberra, eliciting public comment and concern. The long-term changes appraised by Green et al. (2021) suggested that moth numbers were reasonably stable from 1951 (the start of records) until about 1980, fluctuated with slow declines until 2016, and then declined to crash from 2017.

Braby (2019) nominated four possible approaches that could contribute to remedying this lack of information, focused on addressing three key questions. These questions are (1) are Australian insects and other invertebrates in rapid decline, and to what extent are they threatened with extinction?; (2) if they are in decline, are the drivers or key threatening processes similar to those in the northern hemisphere?; and (3) given those threats, what can be done to rectify the situation and reverse the decline? The investigative themes nominated by Braby to stimulate further research are (1) estimating the extent and rate of extinction; (2) estimating the extent of change, especially decline, based on expert opinion; (3) estimating the extent of change based on historical records; and (4) estimating the extent of change based on temporal replication of surveys. Essentially, as noted also by Raven and Yeates (2007), conservation actions for insects must be based on the best information available at that time, and modified as more information accrues. The four themes they nominated as necessary for sustaining Australia's biodiversity (description and understanding, direct action for conservation, promoting sustainable use in conjunction with conservation, providing education and information to maintain this effort) remain foundations for advance. Although complete characterisation and description of Australia's insects and their ecological interactions is still far off, Raven and Yeates' guiding comment that 'the more we know the better choices we shall be able to make' is incontrovertible.

In larger and better-known insect groups, a considerable variety of responses, from increase to total loss, have been reported (Wagner et al. 2021). As might be expected, most conservation attention has been directed to those taxa that are demonstrably 'losers', through having declined in range and/or abundance, become threatened or lost, or simply signalled as of concern. However, other species are 'winners' that become more abundant or more widespread—so that changes in insect faunas includer 'losers', 'winners' and those species that do not undergo obvious changes and appear relatively stable. The more obvious 'winners' include some agricultural herbivores and pests that have burgeoned as suitable crop-scale food has proliferated (Sanchez-Bayo and Wyckhuys 2021). It has been believed widely that 'losers' tend to outnumber 'winners', but an extensive overview (Dornelas et al. 2019) implied that frequency of the two categories in assemblages was approximately balanced for changes in abundance and distribution (occupancy), leading to recommendation that the two need to be considered together in assessing biodiversity change. For British moths, Boyes et al. (2019) used the extensive long-term records available to assess the changes in abundance and distribution of 51 species they considered 'winners', and confirmed that these were ecologically diverse. They included both habitat specialists and more widespread generalist species (in part refuting the common proposition that losers are specialists and winners are generalists, a finding shared by Coulthard et al. 2019), as well as recent insect colonists from continental Europe. Patterns of change in distribution and range over about five decades were frequently non-linear and the extent of changes varied considerably. A considerable range of 'drivers' for change were suggested: climate changes may have facilitated range changes and greater populations for some species, habitat changes could create novel habitats and improve conditions needed by some species, novel genotypes might be introduced from migrants from Europe, and rapid adaptations to anthropogenic changes could occur. A broader survey of changes in moth faunas suggested that the most extensive and severe declines of larger moths have been mostly from the northern hemisphere, predominantly in regions with high human density and intensive agriculture, where a wide range of different stressors occur (Raven and Wagner 2021; Wagner et al. 2021). Boyes et al. (2019) also listed the indirect benefits from habitat management in place for any threatened species as measures that could also benefit other taxa, and some of the trends may not even be related to anthropogenic changes but be a continuation of far longer-term evolutionary trajectories. Finally, they noted possible effects of 'network disruptions' whereby positive effects on winners could be linked with negative trends in losers as competitors and natural enemies are lost. The modelling approach used by Coulthard et al. for British moths explored a number of traits, such as size (wingspan), activity (crepuscular or nocturnal), the primary habitat occupied, and larval food as linking with the proportions of species increasing or decreasing, as summarised in Fig. 1.2. That variety of possible drivers, and uncertainties over their relative impacts on different moth species in even that intensively recorded insect group in an area where such changes can be regarded as valid, is a sobering reminder of how little comparable detailed knowledge is available for any Australian insects. As Rix et al. (2017) wrote in discussing detections of declines in Australian spiders 'By definition, a poorly known fauna lacks rigorous empirical or quantitative assessment at the species and population levels, and this, in turn, renders such groups difficult to allocate from a conservation perspective'. That comment applies equally as a major stricture to understanding in detail the needs of numerous groups of Australian insects, amongst which levels or recent losses or declines can often only be inferred. Whilst Braby et al. (2021) noted four potential sources of assessing declines, namely historical records, expert opinion, survey replication and documented extinctions, evidence is often somewhat tentative beyond site or local landscape levels. Whilst extinctions can be—and are—suggested, for many species some doubt remains because of incomplete surveys and knowledge of the taxon's biology. Minute isolated populations of some may persist in tiny remnant habitats. Referring to insect extinction, Kitching (2004) noted '... it would be a brave entomologist (in Australia at least) who would claim such status for his or her subjects'. There is, however, no room for complacency—with global suggestions of 250,000–500,000 insect extinctions likely since the industrial era commenced, with insects comprising perhaps half the million or so species facing extinction in coming decades (Regnier et al. 2015; IPBES 2019). The warnings from actual or suspected declines cannot be ignored.

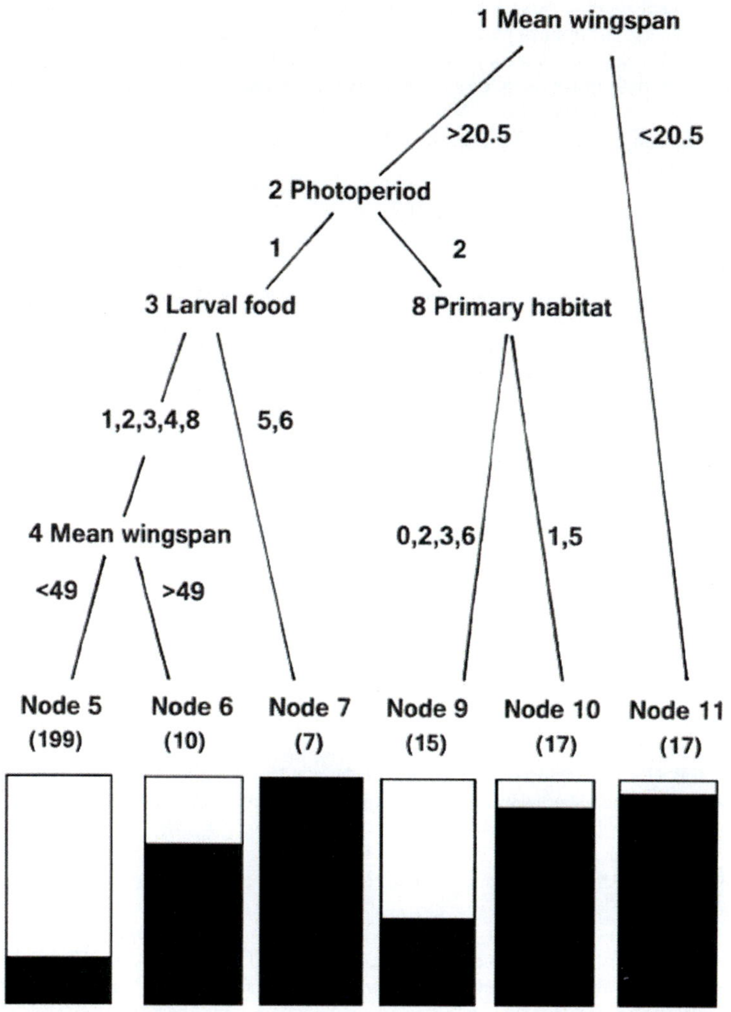

Fig. 1.2 A decision tree showing population changes of moth species according to their ecological and morphological traits for 'common and widespread' British moths over 40 years. Within the tree, the numbered nodes refer to wingspan (size in mm), photoperiod (1, nocturnal; 2, crepuscular), larval food (1, herbaceous; 2, woody plants) and primary habitat (0, no association; 1, woodland; 2, grassland; 3, heathland; 4, urban; 5, coastal; 6, wetland). For each node, the bars show the number of species and the proportions of those species that are increasing (black) or decreasing (open) (after Coulthard et al. 2019)

References

Asher J, Warren M, Fox R, Harding P, Jeffcoate G, Jeffcoate S (2001) Millenium atlas of butterflies in Britain and Ireland. Oxford University Press, Oxford

Basset Y, Lamarre GPA (2019) Toward a world that values insects. Science 364:1230–1231

Basset Y, Samuelson GA, Allison A, Miller SE (1996) How many species of host-specific insect feed on a species of tropical tree? Biol J Linn Soc 59:201–216

Beirne BP (1955) Natural fluctuations in abundance of British Lepidoptera. Entomol Gaz 6:21–52

Berenbaum M (2009) Insect biodiversity – millions and millions. In: Foottit R, Adler P (eds) Insect biodiversity: science and society. Wiley-Blackwell, Oxford, pp 575–582

Borkent A, Brown BV, Adler PH, De Souza AD, Barber K et al (2018) Remarkable fly (Diptera) diversity in a patch of Costa Rican cloud forest: why inventory is a vital science. Zootaxa 4402: 53–90

Boyes DH, Fox R, Shortall CR, Whittaker RJ (2019) Bucking the trend: the diversity of Anthropocene 'winners' among British moths. Front Biogeogr 11(3):e4386.2. https://doi.org/10.21425/FSFBG43862

Braby MF (2019) Are insects and other invertebrates in decline in Australia? Austral Entomol 58: 471–477

Braby MF, Williams MR, Douglas F, Beardsell C, Crosby DF (2021) Changes in a peri-urban butterfly assemblage over 80 years near Melbourne, Australia. Austral Entomol 60:27–51

Brown BV, Borkent A, Adler PH, De Souza AD, Barber K et al (2018) Comprehensive inventory of true flies (Diptera) at a tropical site. Commun Biol 1:21. https://doi.org/10.1038/s42003.018-0022.x/www.nature.comm/commsbio

Cameron SA, Lozier JD, Strange JP, Loch JB, Cordes N, Solter LF, Griswold TL (2011) Patterns of widespread decline in north American bumble bees. Proc Nat Acad Sci 108:662–667

Cardoso P, Leather SR (2019) Predicting a global insect apocalypse. Insect Conserv Divers 12:263–267

Cardoso P, Barton PS, Birkhofer K, Chichorro F, Deacon C et al (2020) Scientists' warning to humanity on insect extinctions. Biol Conserv 242:108426

Cassis G (1999) Workshop: putting invertebrates on the agenda: opportunities for the future. In: Ponder W, Lunney D (eds) The other 99%. The conservation and biodiversity of invertebrates. Royal Zoological Society of New South Wales, Mosman, pp 437–445

Coleman DC, Hendrix PF (eds) (2000) Invertebrates as webmasters in ecosystems. CABI, Wallingford

Common IFB (1954) A study of the ecology of the adult Bogong moth, *Agrotis infusa* (Boisd.) (Lepidoptera: Noctuidae), with special reference to its behaviour and aestivation. Aust J Zool 2: 223–263

Conrad KF, Warren M, Fox R, Parsons M, Woiwod IP (2006) Rapid declines of common, widespread British moths provide evidence of an insect biodiversity crisis. Biol Conserv 132: 279–291

Cooney SJN, Olsen PD, Garnett ST (2009) Ecology of the coprophagous moth *Trisyntopa neossophila* Edwards (Lepidoptera: Oecophoridae). Aust J Entomol 48:97–101

Coulthard E, Norrey J, Shortall C, Harris WE (2019) Ecological traits predict population changes in moths. Biol Conserv 233:213–219

Dasgupta P (2021) The economics of biodiversity: the Dasgupta review. HM Treasury, London

Didham RK, Basset Y, Collins CM, Leather SR, Littlewood NA et al (2020) Interpreting insect declines: seven challenges and a way forward. Insect Conserv Divers 13:103–114

Dornelas M, Gotelli NJ, Shimadzu H, Moyes F, Magurran AE, McGill BJ (2019) A balance of winners and losers in the Anthropocene. Ecol Lett 22:847–854

Dougherty ER, Carlson CJ, Bueno VM, Burgio KR, Cizauskas CA et al (2016) Paradigms for parasite conservation. Conserv Biol 30:724–733

Duffus NE, Christie CR, Morimoto J (2021) Insect cultural services: how insects have changed our lives and how can we do better for them. Insects 12:377. https://doi.org/10.3390/imsects12050377

Dunn RR (2005) Modern insect extinctions, the neglected majority. Conserv Biol 19:1030–1036

Edwards ED, Cooney SJN, Olsen P, Garnett ST (2007) A new species of *Trisyntopa* Lower (Lepidoptera: Oecophoridae) associated with the nests of the hooded parrot (*Psephotus dissimilis*, Psittacidae) in the Northern Territory. Aust J Entomol 46:276–280

Erwin TL (1982) Tropical forests: their richness in Coleoptera and other arthropod species. Coleopt Bull 36:74–75

Essens T, van Langevelde F, Vos RA, van Swaay CAM, Wallis DeVries MF (2017) Ecological determinants of butterfly vulnerability across the European continent. J Insect Conserv 21:439–450

Flood J (1980) The moth hunters: aboriginal prehistory of the Australian alps. Australian Institute of Aboriginal Studies, Canberra

Fonseca CR (2009) The silent mass extinction of insect herbivores in biodiversity hotspots. Conserv Biol 23:1507–1515

Forbes AA, Bagley RK, Beer MA, Hippee AC, Widmayer HA (2018) Quantifying the unquantifiable: why Hymenoptera, not Coleoptera, is the most speciose animal order. BMC Ecol 18:21. https://doi.org/10.1186/s12898-018-0176-x

Forister ML, Pelton EM, Black SH (2019) Declines in insect abundance and diversity: we know enough to act now. Conserv Sci Pract 1:e80

Fox R (2013) The decline of moths in Great Britain: a review of possible causes. Insect Conserv Divers 6:5–19

Fox R, Bourn NAD, Dennis EB, Heafield RT, Maclean IMD, Wilson RJ (2019) Opinions of citizen scientists on open access to UK butterfly and moth occurrence data. Biodivers Conserv 28: 3321–3341

Garcia-Robledo C, Kuprewicz EK, Baer CS, Clifton E, Hernandez GG, Wagner DL (2020) The Erwin equation of biodiversity: from little steps to quantum leaps in the discovery of tropical insect diversity. Biotropica 52:590–597

Gaston KJ, Hudson E (1994) Regional patterns of diversity and estimates of global insect species richness. Biodiv Conserv 3:493–500

Goulson D, Lye GC, Darvill B (2008) Decline and conservation of bumble bees. Annu Rev Entomol 53:191–208

Green K, Caley P, Baker M, Dreyer D, Wallace J, Warrant E (2021) Australian Bogong moths *Agrotis infusa* (Lepidoptera: Noctuidae), 1951-2020: decline and crash. Austral Entomol 60:66–81

Groenendijk D, Ellis WN (2011) The state of the Dutch larger moth fauna. J Insect Conserv 15:95–101

Habel JC, Segerer A, Ulrich W, Torchyk O, Weisser WW, Schmitt T (2016) Butterfly community shifts over two centuries. Conserv Biol 30:754–762

Habel JC, Samways MJ, Schmitt T (2019) Mitigating the precipitous decline of terrestrial European insects: requirements for a new strategy. Biodivers Conserv 28:1343–1360

Hallmann CA, Sorg M, Jongejans E, Siepel H, Hofland N et al (2017) More than 75 percent decline over 27 years in total flying insect biomass in protected areas. PLoS One 12:e0185809

Hallmann CA, Zeegers T, van Klink R, Vermuelen R, van Wielink P et al (2020) Declining abundance of beetles, moths and caddisflies in the Netherlands. Insect Conserv Divers 13: 127–139

Harvey JA, Heinen R, Armbrecht I, Basset Y, Baxter-Gilbert JH et al (2020) International scientists formulate a roadmap for insect conservation and recovery. Nat Ecol Evol 4:174–176

Haslett JR (2008) European strategy for the conservation of invertebrates. Nature and Environment, no 145. Council of Europe Publishing, Strasbourg

Hebert PDN, Penton EH, Burns JM, Janzen DH, Hallwachs W (2004) Ten species in one: DNA barcoding reveals cryptic species in the neotropical skipper butterfly *Astraptes fulgerator*. Proc Nat Acad Sci 101:14812–14817

Hebert PND, Ratnasingham S, Zakharov EV, Telfer AC, Levesque-Beaudin V et al (2016) Counting animal species with DNA barcodes. Canadian insects. Philos Trans R Soc Lond B371(1702):20150333

Hochkirch A (2016) The insect crisis we can't ignore. Nature 539:141

IPBES (2019) Summary for policymakers of the global assessment report on biodiversity and ecosystem Services of the Intergovernmental Science-Policy Platform on biodiversity and ecosystem services. IPBES Secretariat, Bonn

Janzen DH (1994) Wildland diversity management in the tropics: where are we now, and where are we going? Vida Silvestre Neotropical 3:3–15

Janzen DH (1997) Wildland biodiversity management in the tropics. In: Reaka-Kudla ML, Wilson DE, Wilson EO (eds) Biodiversity II. Understanding and protecting our biological resources. Joseph Henry Press, Washington, DC, pp 411–431

Janzen DH, Hallwachs W (2016) DNA barcoding the Lepidoptera inventory of a large complex tropical conserved wildland, Area de Conservacion Guanacaste, norhwestern Costa Rica. Genome 59:641–660

Janzen DH, Hallwachs W (2019) Perspective: where might be many tropical insects? Biol Conserv 233:102–108

Kehoe R, Frago D, Sanders D (2020, 2020) Cascading extinctions as a hidden driver of insect decline. Ecol Entomol. https://doi.org/10.1111/een.12985

Kitching RL (2004) Invertebrate conservation and the conservation of forests. In: Lunney D (ed) Conservation of Australia's forest fauna, 2nd edn. Royal Zoological Society of New South Wales, Mosman, pp 115–126

Komonen A, Halme P, Kotiaho JS (2019) Alarmist by bad design: strongly popularized unsubstantiated claims undermine credibility of conservation science. Rethink Ecol 4:17–19

Kosior A, Celary W, Oejniczak P, Fijal J, Krol W, Soarz W, Plonka P (2007) The decline of the bumble bees and cuckoo bees (Hymenoptera: Apidae: Bombini) of Western and Central Europe. Oryx 41:79–88

Kwak M (2018) Australia's vanishing fleas (Insecta: Siphonaptera): a case study in methods for the assessment and conservation of threatened flea species. J Insect Conserv 22:545–550

Larsen BB, Miller EC, Rhodes ME, Wiens JJ (2017) Inordinate fondness multiplied and redistributed: the number of species on earth and the new pie of life. Q Rev Biol 92. https://doi.org/10.1086/693564

Laussmann T, Dahl A, Radtke A (2021) Lost and found: 160 years of Lepidoptera observations in Wuppertal (Germany). J Insect Conserv 25:273–285

Leandro C, Jay-Robert P, Vergnes A (2017) Bias and perspectives in insect conservation: a European scale analysis. Biol Conserv 215:213–224

Lister BC, Garcia A (2018) Climate-driven declines in arthropod abundance restructure a rainforest food web. Proc Natl Acad Sci U S A 115:E10397–E10406

Mattila M, Kaitala V, Komonen A, Kotiaho JS, Paivinen J (2006) Ecological determinants of distribution declines and risk of extinction in moths. Conserv Biol 20:1161–1168

Mittermeier RA, Robles-Gil P, Hoffman M, Pilgrim J, Brooks T et al (2004) Hotspots revisited: Earth's biologically richest and most endangered terrestrial ecoregions. University of Chicago Press, Chicago

Moir ML (2021) Coextinction of *Pseudococcus markharveyi* (Hemiptera: Pseudococcidae): a case study in the modern insect extinction crisis. Austral Entomol 60:89–97

Montgomery GA, Dunn RR, Fox R, Jongejans E, Leather SR et al (2020) Is the insect apocalypse upon us? How to find out. Biol Conserv 241:108327

Owens ACS, Cochard P, Durrant J, Farnworth B, Perkin EK, Seymoure B (2020) Light pollution is a driver of insect declines. Biol Conserv 241:008259

Potts SG, Imperatriz-Fonseca I, Ngo HT, Aizen MA, Biesmeijer JC et al (2016) Safeguarding pollinators and their values to human well-being. Nature 540:220–229

Raven PH, Wagner DL (2021) Agricultural intensification and climate change are rapidly decreasing insect biodiversity. Proc Nat Acad Sci 118(2):e2002548117

Raven PH, Yeates DK (2007) Australian biodiversity: threats for the present, opportunities for the future. Aust J Entomol 46:177–187

Regnier C, Achaz G, Lambert A, Cowle RH, Bouchet P, Fontaine B (2015) Mass extinction in poorly known taxa. Proc Nat Acad Sci 112:7761–7766

Rix MG, Huey JA, Main BY, Waldock JM, Harrison SE et al (2017) Where have all the spiders gone? The decline of a poorly known invertebrate fauna in the agricultural and arid zones of southern Australia. Austral Entomol 56:14–22

Rosza L, Vas Z (2014) Co-extinct and critically co-endangered species of parasitic lice, and conservation-induced extinction: should lice be reintroduced to their hosts? Oryx 49:107–110
Samways MJ (2020) Insect conservation. A global synthesis. CABI, Wallingford
Sanchez-Bayo F, Wyckhuys KAG (2019) Worldwide decline of the entomofauna: a review of its drivers. Biol Conserv 232:8–27
Sanchez-Bayo F, Wyckhuys KAG (2021) Further evidence for a global decline of the entomofauna. Austral Entomol 60:9–26
Saunders ME (2019) No simple answers for insect conservation: media hype has missed the biggest concern that ecologists and entomologists have about six-legged life: how little we know about it. Am Sci 107(3):148
Saunders ME, Janes JK, O'Hanlon JC (2020a) Semantics of the insect decline narrative: recommendations for communicating insect conservation to peer and public audiences. Insect Conserv Divers 13:211–213
Saunders ME, Janes JK, O'Hanlon JC (2020b) Moving on from the insect apocalypse narrative: engaging with evidence-based insect conservation. Bioscience 70:80–89
Schowalter TD, Willig MR, Presley SJ, Pandey M (2019) Warnings of an "insect apocalypse" are premature. Front Ecol Environ 17:547
Shaw MR, Hochberg ME (2001) The neglect of parasitic Hymenoptera in insect conservation strategies: the British fauna as a prime example. J Insect Conserv 5:253–263
Simmons BI, Balmford A, Bladon AJ, Christie AP, De Palma A et al (2019) Worldwide insect declines: an important message, but interpret with caution. Ecol Evol 9:3678–3680
Stork NE (2018) How many species of insects and other terrestrial arthropods are there on earth? Annu Rev Entomol 63:31–45
Stringer AP, Linklater W (2014) Everything in moderation: principles of parasite control for wildlife conservation. Bioscience 64:932–937
Thomas CD, Jones TH, Hartley SE (2019) "Insectageddon": a call for more robust data and rigorous analyses. Glob Change Biol 25:1891–1892
Valentine PS (2004) The demise of mass migration of the Brown Awl *Badamia exclamationis* (Fabricius 1775) (Lepidoptera: Hesperiidae): a consequence of land clearing in Queensland? Pac Conserv Biol 10:67–69
van der Sluijs JP (2019) Insect decline, an emerging global environmental risk. Curr Opin Environ Sustain 20:1–4
van Klink R, Bower DE, Gonglalsky KB, Swengel AB, Gentile A, Chase JM (2020) Meta-analysis reveals declines in terrestrial but increases in freshwater insect abundances. Science 368:417–420
Vogel G (2017) Where have all the insects gone? Science 356:576–579
Wagner DL (2020) Insect declines in the Anthropocene. Annu Rev Entomol 65:457–480
Wagner DL, Fox R, Salcido DM, Dyer LA (2021) A window to the world of global insect declines: moth biodiversity trends are complex and heterogeneous. Proc Nat Acad Sci 118(2): e2002549117
Williams PH (1989) Bumble bees – and their decline in Britain. Central Association of Bee-Keepers, Ilford
Wilson EO (1987) The little things that run the world (the importance and conservation of invertebrates). Conserv Biol 1:344–346
Wilson RJ, Fox R (2020) Insect responses to global change offer signposts for biodiversity and conservation. Ecol Entomol. https://doi.org/10.1111/een.12970
Woinarski JC, Burbidge AA, Harrison PL (2015) Ongoing unraveling of a continental fauna: decline and extinction of Australian mammals since European settlement. Proc Nat Acad Sci 112:4531–4540
Woinarski J, Braby MF, Burbidge AA, Coates D, Garnett ST et al (2019) Reading the black book: the number, timing, distribution and causes of listed extinctions in Australia. Biol Conserv 239: 108261
Yen AL, Butcher RJ (1997) An overview of the conservation of non-marine invertebrates in Australia. Environment Australia, Canberra

Chapter 2
The Problems of Interpreting Changes

2.1 Introduction

The natural fluctuations in many insect populations can occur rapidly (so that numbers of individuals in successive generations can differ substantially) or be more predictable in following a more cyclic pattern of increase and decrease over longer periods. In most cases, any such pattern is unknown and the causes of changes, some of which are unpredictable, are also unknown in detail. At extremes, a few taxa undergo sporadic outbreaks, when their normally low abundance increases massively to reach enormous numbers and, in some cases, cause economic damage. In Australia, such outbreaks are of particular concern among some forest stick insects that can then defoliate eucalyptus forests in the south-east, and a few species of grasshoppers that become swarming 'locusts' on crops and grasslands. Causes of such phenomena have been investigated in considerable detail, but many uncertainties remain, with implications of weather, migrations and other influences on insect abundance.

Likewise, the composition and richness of insect communities is dynamic, with species continuously entering or leaving a community or site, either participating in its functions and interacting with other species or as more casual 'tourists' that are not resident as permanent breeding populations. A further dilemma is simply that it is usually unclear what population structure the species under study manifests. Many insects have a metapopulation structure, in which small populations (often those of greatest initial concern to well-intentioned conservationists, as deemed vulnerable) on isolated habitat patches across a landscape may naturally die out, to be replaced by later colonisations as part of their normal population pattern—in which local extinctions are entirely natural and for which conservation devolves on the maintenance of suitable habitat patches within the insect's dispersal range and that may at some time participate in maintaining the population within the wider area.

Unlike the relatively well-known insect faunas of parts of the northern hemisphere, assessing fully the fauna of even a small region of Australia from adequate

© The Author(s), under exclusive license to Springer Nature Switzerland AG 2022

T. R. New, *Insect Diversity, Declines and Conservation in Australia*, Fascinating Life Sciences, https://doi.org/10.1007/978-3-030-90134-9_2

samples is extremely difficult, and the general lack of detailed historical information in forms that can be readily verified renders detection of changes rather tentative. A few cases exist in which detailed field notebooks from earlier explorations facilitate visits to the same sites explored previously to compare the fauna present across the two surveys, together with changes in habitat condition. Perhaps the most significant and carefully explored example is the resurvey in 2019 of a series of grasshopper collecting sites in New South Wales, as part of extensive surveys by Key in the 1940s and described carefully in a series of notebooks deposited in the Australian National Insect Collection, Canberra (Kearney et al. 2021). In all, more than 220 such field notebooks record Key's extensive field surveys throughout Australia, accompanying the extensive series of specimens collected. Using the odometer readings and site descriptions in Key's notes, Kearney et al. revisited 45 sites in rural New South Wales, and collected grasshoppers using the same timed surveys as a basis. Many of the areas had changed little in the intervening 70 years, although some showed increased weedy grass and shrub presence. In general, more grasshoppers were found in the recent surveys, but incomplete taxonomy prevented adult identifications beyond genus-level and nymphs could not be diagnosed reliably. This survey confirmed the values of revisiting older information in this way, but also the continuing lack of full interpretations because of incomplete taxonomic knowledge.

Refining any baseline against which to measure changed abundance, distribution and even incidence of particular insect taxa, and describing and evaluating species richness within communities, raises series of practical problems but is only part of the challenges that face conservation biologists and specialist entomologists alike. In many regional faunas, simply detecting 'declines' in terms of the numbers of species present is difficult and even for Europe and North America any such number is continuously dynamic as additional discoveries are made, taxonomic revisions are accomplished (in which, typically, new species are described and some previously thought distinct are synonymised, so that the number of species is 'edited' by both additions and subtractions) and the fauna of under-explored regions documented more completely. Almost universally, however, insects are the most species-rich and abundant animals in all habitable terrestrial and freshwater ecosystems. Their varied roles, through participating in food webs and other ecological processes at all trophic levels, sustain many of those ecosystems and the diversity of other organisms present. Understanding how and why these major components of biodiversity are influenced—whether a species' abundance is declining, increasing or staying in balance, or whether changes in biomass or community richness occur—by disturbances or imposed changes contributes to how the natural world can be protected and managed. This chapter includes a summary of several important themes, some commonly overlooked, inherent in trying to document and understand insect diversity and abundance. Changes in insect populations or communities have a number of discrete dimensions, and each may be the primary focus of a survey. These dimensions can be categorised broadly into four groups (Montgomery et al. 2021), as (1) changes in occurrences and distribution; (2) phenology, as shifts in seasonal development or appearance; (3) abundance and biomass; and (4) species richness and species composition as measures of biodiversity. The first and last of these are the most frequently emphasised, with increasing attention to phenology occurring as studies on climate change influences proliferate and potentially affect the seasonal synchrony between insect consumers and their foods (Moir et al. 2014).

As noted above, detecting changes in either a species' abundance or wider species richness requires standardised and replicated sampling regimes, preferably over extended periods to 'iron out' normal (but almost always unknown) fluctuations, and other than a very few long-term regular monitoring programmes (p. 48), are difficult to undertake. Without monitoring over sufficiently long periods to incorporate the largely unknown 'natural' fluctuations in abundance, or without understanding the population structure of the target species, trends of decline can usually only be inferred—or firmer declarations of losses can become open to query. Seasonal standardisation is also necessary. Many insect species are amenable to sampling only over a short period each year, and the precise flight season of many butterflies, for example, can be influenced by temperature over the previous months. Recent concerns over bee declines have led to attempts to design suitable surveys by which to evaluate changes. LeBuhn et al. (2013), for example, outlined a programme that involved monitoring of 200–250 locations, each to be sampled twice every 5 years by pan traps and projected as capable of detecting small (2–5%) annual declines in richness and abundance. Aspects of the proposal were criticised (Tepedino et al. 2015, and response by LeBuhn et al. 2015), with a major concern stated being that 'as presented, the authors' [i.e. Lebuhn et al., my addition] protocol cannot identify locations, habitats or species at risk; illuminate the case of a significant bee decline; or provide a remedy'. Tepedino et al., rather, urged surveillance studies in more specific contexts involving specific threats in selected areas, from which outcomes could include (1) where bee populations are declining; (2) the factors associated with those declines; and (3) develop approaches to reverse the trend. In short, detecting any declines can become a complex logistic exercise with widely differing interpretations of any changes found.

The intractability of some of the themes discussed below contributes in large part to the relative neglect of insects (and other invertebrates) in practical conservation, where they are often either (1) dismissed as unimportant or (2) considered 'too hard' to incorporate into wider conservation agendas or treat as primary conservation targets. Collectively, the relevant themes pose massive challenges, but also opportunities in emphasising the current lacunae in ability to enumerate and assess conservation needs in all but the very best-documented taxa, and through which credibility in processes such as (1) selection and nomination of taxa for some form of threatened species listing, (2) estimates of species richness and assemblage diversity and (3) evaluating impacts of environmental change can be increased. They also demonstrate the vast gaps in understanding faced in improving this reward through more reliable and more complete information.

A series of seven major 'impediments' (Cardoso et al. 2011a) encapsulated many of the relevant themes. They are introduced here, and their relevance to the Australian fauna expanded in Chap. 6. Listed in Table 2.1, this framework has been amplified over the last decade, and recapitulates many of the practical and

Table 2.1 The seven impediments in invertebrate conservation, as identified by Cardoso et al. (2011a) (see text)

1. Invertebrates and their ecological services are mostly unknown to the general public (the public dilemma)
2. Policy-makers and stakeholders are mostly unaware of invertebrate conservation problems (the political dilemma)
3. Basic science on invertebrates is scarce and underfunded (the scientific dilemma)
4. Most species are undescribed (the Linnean shortfall)
5. The distribution of described species is usually unknown (the Wallacean shortfall)
6. The abundance of species and their changes in space and time are unknown (the Prestonian shortfall)
7. Species' ways of life and sensitivities to habitat change are largely unknown (the Hutchinsonian shortfall)

theoretical problems of understanding conservation status and needs—and also why invertebrates have received so little attention in conservation. The seven impediments fall into two rather different groups. Four are 'scientific', founded in 'knowledge', and are the primary concerns here in defining areas of knowledge reflecting lack of understanding, information and thwarting ability to make rational and optimal conservation decisions and progress. The other three are 'societal', as themes addressed in taking decisions on how important invertebrate conservation may be and are founded in 'insect values'. Reflecting these different emphases, the societal themes were designated 'dilemmas' and the scientific themes as 'shortfalls'. Their relationships are simplistically summarised in Fig. 2.1, and comments on each follow below.

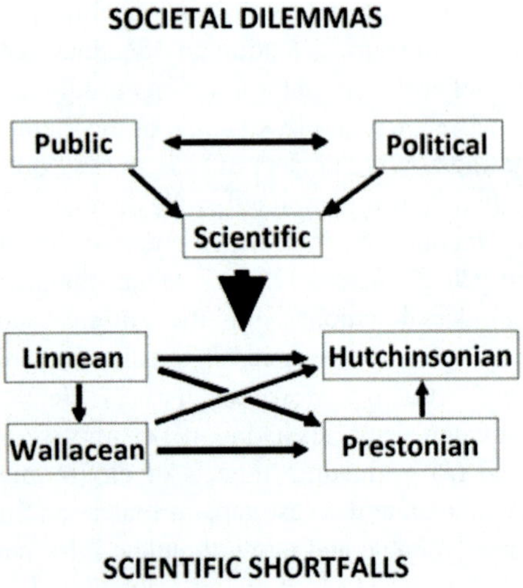

Fig. 2.1 The seven impediments in invertebrate conservation and relationships between them (Cardoso et al. 2011a)

2.2 Scientific Shortfalls

1. The Linnean shortfall, that most species are undescribed. Entomologists continue to debate how many insect species occur on Earth, but generally agree that there are a lot! A likely range of 5–10 million species is accepted widely, although some earlier estimates were far higher, up to 30 million or more. Much depends on how 'species' are delimited and recognised, and without that agreement, different experts will continue to advance rather different numbers even within the same higher taxon. Whatever the number/s adopted, it is clear that many species await recognition, definition, description and even collecting. Even in western Europe, and North America, taxonomic reviews continue to reveal previously unrecognised taxa—and also to discover or validate synonymies so that even in such well-documented faunas and taxonomic groups any fine-tuned total is a dynamically edited number as investigations continue, new species are described and some previously described ones removed from the tally, to reflect a shifting baseline of numbers and species boundaries. Some major insect groups remain uncensussed in even the best-known areas. Small parasitoid wasps are a commonly-cited example (Shaw and Hochberg 2001). Numerous species are still very poorly known even within the United Kingdom's approximately 27,000 insect species (Eggleton 2020) as a fauna studied over '300 years of work for many hundreds of taxonomists'. More generally on insects, Eggleton opined that 'we have described no more than some 20% of the species that are known to exist'. And, in estimating numbers that 'Perhaps insects are just too diverse . . . for generalisations to be made'.

   The 'taxonomic impediment', a descriptor originating from Taylor (1976,1983), reflects that poverty of documentation and understanding of richness. Especially in the rich insect faunas of the tropics, undescribed and previously unknown species are encountered frequently—and it can become somewhat unusual to encounter clearly described species in some groups! Consistent recognition of described species is difficult when their patterns of variation are unknown, and for many taxa, accurate identification can be made only by a few specialists aware of recent developments and publications. Even then, species numbers recorded from the most detailed diagnoses from morphological study, as the traditional taxonomic tool for differentiating species, may be changed vastly by different approaches such as DNA barcoding, from which far greater apparent species numbers may be revealed. Temptations by less-experienced workers in the frequent situations where such advice is not available readily to use published accounts to identify insects or use the 'best-available' name they encounter may lead to inaccuracy and dissemination of false information that can mislead conservation and other decisions. Related taxonomic problems include (1) failure to recognise undescribed species that are 'forced' to a name from an outdated key; (2) unpublished synonymies, so that unnecessary names are perpetuated; (3) incidences of intraspecific variation, such as sexual dimorphism, seasonal variation, and minor individual variations in colour or size; (4) difficulties of associating larval and adult stages; and (5) simple 'mistakes' from inexperience or failure to interpret morphological characters accurately. Standards for taxonomic descriptions have changed vastly since the early practice of simply a few lines, with or without illustration, of major morphological features without regard for patterns of variation and well before use of DNA and related diagnostic tools became available. Reliance on those descriptions alone for identification can be highly misleading—writing on mites, but applicable to numerous insect groups, Beard et al. (2015) commented that early descriptions 'are plagued with misidentifications, inconsistencies, erroneous or incomplete drawings, ignored characters, and inadequate host plant identifications'.

Association of early and adult stages may be particularly pertinent when enumerating species from field samples, and the larvae and adults of many species can be very different in appearance and occupy different habitats. For numerous taxa, they have not been associated directly by rearing. The larvae of holometabolous insects do not resemble the adult appearance (such as caterpillar and butterfly), and the secondary adaptations to adopt an aquatic lifestyle in many hemimetabolous insects give larvae very characteristic appearance and features. Knowledge of the latter has considerable practical importance, as those larvae are encountered abundantly in freshwater surveys and are used extensively as 'indicators' of water and habitat quality. Most commonly, larvae of mayflies (Ephemeroptera), stoneflies (Plecoptera) and caddisflies (Trichoptera) are employed together as a basic comparative index of environmental condition, as the 'EPT Index', with dragonflies (Odonata) sometimes added as a fourth metric group (so, EPTO). For these orders, in contrast to many terrestrial insects, keys to larval stages may become as useful, or more relevant to ecological analysis, than those to adults in evaluating diversity.

Current rates of taxonomic descriptions of insect species imply that Earth's insect fauna will never be fully documented, even at the most basic level of naming the more obviously recognisable species. Even for the most conservative estimates Cardoso et al. (2011a) suggested that around a century of effort at current rates will be needed to complete the process from the theoretical base that half the species are already described. It is almost inevitable that hundreds of thousands of species will become extinct before they can be described, or even detected or collected (Gonzalez-Oreja 2008). That theme links closely with the scientific dilemma (p. 40). A frequently repeated maxim in conservation is that in order to conserve 'biodiversity' we need to know what it is: for many insects that basic presumption cannot be fulfilled at present, and conservation must proceed without complete definition of its targets. In Australia, 'land managers currently struggle to determine which insect species inhabit their lands, much less identify which are in need of conservation' (Moir 2015).

2. The Wallacean shortfall, that the distribution of most known insect species is unknown in detail. Gaining a binomial name or other specific recognition label is commonly based on examination of dead specimens, accompanied by minimal data—conventionally the place and date of collection, and collector's name, and usually with only minimal (if any) biological data, so that the total pool of

information is that on the species' specimen labels, which in many older specimens is very curt. Vast numbers of insect species are known from very few specimens, many of them from singletons, from a single site (for many, poorly described—many older specimen labels simply note the country or other large area of origin), collected decades ago and which have not been collected subsequently. The real distribution of such species in the absence of additional surveys and attempts to find them is simply conjecture, perhaps 'informed' by better knowledge of putatively similar taxa that are better understood.

The name 'Wallacean' refers to Alfred Russel Wallace, whose pioneering studies on patterns of geographical distribution founded interests in 'where species occur'. The shortfall emphasises that sound distributional data is vital in conservation planning, simply to evaluate species' status and set priorities over where conservation should be focused.

Many insects have very circumscribed and restricted distributions within a region or country, and some level of endemism (p. 108) is usual. Whilst 'narrow-range endemism', occurrence over a very small range, is common and in many cases supported by targeted searches elsewhere failing to discover the species, any contractions of range due to disturbances or more natural changes such as vegetation succession or climate change may cause concern. For such species, severe disturbance over only very small areas can be disastrous.

3. The Prestonian shortfall, that the changes of abundance and dynamics of most species are poorly or entirely understood in time and space. Most samples of insect faunas from which our knowledge of taxonomy and distribution has been derived represent either (1) 'single occasion' collections, often very restricted in breadth of coverage, with only particular focal taxa retained and as unspecified subsets of any total catch considered, and are commonly from a single locality or few sites; or (2) interval samples, such as seasonal or yearly collections at the same localities and using the same methods, from which suggestions of numerical or representational change can be derived. The epithet 'Prestonian' refers to the ecologist Frank W. Preston, whose work on relative abundances (commonness and rarity) of species and their changes in time and space highlighted the importance of recognising those changes, their causes, and their wider ecological correlates.

   Applications of the standardised sampling regimes needed to plot insect numbers in space and time can in principle provide massive amounts of information. However, in almost all cases with priority for inventory and monitoring, much of that information is never harnessed. The usual 'snapshot' of an insect species or community from a sample can certainly tell us what was there on that particular occasion and (if full catches are retained for study) provide some suggestion of relative abundance within that single retrieval.

4. The Hutchinsonian Shortfall, that the insects' ways of life and sensitivity to environmental changes are unknown in any detail. The name commemorates G. Evelyn Hutchinson, whose influential pioneering studies on the 'niche concept' emphasised the ways in resources affect the abundance and distribution of species.

Many insects can be allocated broadly and with varying levels of confidence to some trophic group, and some of their key needs (such as host plant species and features) inferred. However, numerous details of autecology and interactions within the host community are generally unknown, so that management can often not be honed to cater for specific requirements that can be revealed only through detailed study.

The broader picture is that almost nothing is known of how particular insects contribute to ecosystem wellbeing or 'ecosystem services' and, so, what the consequences of losing those species may be on its home biotope and community. Likewise, the impacts of external changes can usually not be projected comprehensively.

The last two of these shortfalls are the least heeded and most intangible, with attempts to assess identities and distributions of insect species the foundation of much conservation practice. They are also the basis for conservation status evaluation and, so, the urgency of their conservation needs. That recognition commonly fosters the more extensive ecological work important in the Prestonian and Hutchinsonian shortfalls, both of which emphasise the wider spatial and temporal contexts of insect persistence.

2.3 Societal Dilemmas

The three societal dilemmas reflect the widespread lack of public interest in insects, ignorance of their variety and importance, and the logistic constraints that hamper progress toward addressing the goals of the scientific shortfalls.

1. The public dilemma, that insects and the ecological services they provide are largely unknown to the general public. One of the main reasons that butterflies have such a predominant 'flagship' role in insect conservation is simply that—broadly—people like them or, at least, do not fear them and accept them as in some way enhancing quality of life through their aesthetic appeal. Few other insects gain such widespread sympathy. Whilst media and scientist concerns over losses of insect pollinators have led to much public support for bees (perhaps especially for bumblebees in the northern hemisphere), this is more pragmatic, and relatively few people would claim to like most smaller bees for themselves rather than the services they provide—or even to know that 'bees' applies to far more than the European honey bee and a few other familiar taxa. A few other insects have substantial cultural significance: Odonata differ from all other aquatic insects in that way—as for butterflies, many dragonflies are brightly coloured, day-active and conspicuous, and the two groups are sometimes referred to as 'birdwatchers' bugs'. Most other insects lack any such appeal and to many people all are 'pests' either actually or potentially affecting commodities, health or aesthetic sensibilities. In short, most insect groups are not 'charismatic' and have little chance of becoming so. That public prejudices can severely hamper

advances in insect conservation is acknowledged widely, and more effective communication between scientists and 'the public' is key to overcoming this. The division was illustrated in New Zealand, where Salvador et al. (2021) showed that the interests and priorities of the two parties were 'only tangentially aligned'. Some widespread beliefs were not endorsed by their surveys—for example 12.5% of the reapondent scientists ranked Lepidoptera highest among invertebrate groups likely to fascinate the public (as suggested above), with the Monarch (or Wanderer) butterfly, *Danaus plexippus*, and the European honey bee, *Apis mellifera*, among the most useful species to capture public attention. In contrast, public responses did not reveal Lepidoptera to be particularly interesting, with cnidarians (especially jellyfish) and crustaceans more popular than any insects. A major inference is that far more extensive and effective dialogue is needed to promote values of insects and their conservation.

Thus, the widespread positive roles of many insects in sustaining the natural world are largely overlooked in the public realm, so that grounds for conserving a small threatened moth, beetle or wasp become very difficult to communicate effectively. Without that public support and awareness most of these, and other invertebrates, will continue to be neglected.

Positive publicity, coaxed through unusual features of a species, pride in local endemics, evidence of specific threats, ecological value and even 'simple measures' such as using a catchy common name to increase appreciation, can radically change public opinion and help to introduce aspects of conservation need and management, not least through fostering participation in 'citizen science' (Chap. 9). Common names can reflect local identity and foster local interest, as occurred for two related threatened lycaenid butterflies in southern Australia. The Eltham copper (*Paralucia pyrodiscus lucida*) and the Bathhurst copper (*Paralucia spinifera*) were both named for the places (in Victoria and New South Wales, respectively) where conservation interest in them was founded. Cardoso et al. (2011a) noted that memorable scientific names for species—such as naming them after celebrities—can also increase public interest. For example, Janzen and Hallwachs (2016) noted that the newly described insects from Costa Rica included taxa named after each of the 150 or so parataxonomist helpers (p. 60) involved in their continuing surveys, as well as one for the President of Costa Rica. A spectacular new species of New Zealand's primary flagship insect group, weta (Orthoptera, p. 100) has recently been named after the country's Prime Minister (Trewick 2021).

2. The political dilemma, that many policymakers and stakeholders are largely unaware of insects, and of the diversity and ecological characteristics of insects that render them important to conserve. A widespread assumption, should insects be considered at all, is that there is little need for their specific consideration in conservation because they (and other invertebrates) are 'automatically' safeguarded through measures taken to conserve larger vertebrates. The latter are thus treated as 'umbrella species'. Consequently, funds and logistic support for specific invertebrate conservation programmes are minimal, and sometimes even regarded as diversion or erosion of support from where it 'is really needed'.

However, the above assumption is often simply wishful, without evidence to confirm its truth, however desirable in terms of honing conservation effort by relying on such well-documented taxa as mammals and birds, together with vascular plants. As Schuldt and Assmann (2010) noted, the extent of cross-taxon congruence is reflected strongly in the scale of study: small-scale surveys (such as in individual countries or reserves in Europe) indicate that correspondence between diversity patterns of different invertebrate groups or congruence with vertebrate or plant diversity is often weak. However, the opposite has also been reported.

In many heavily altered landscapes, reserves and other set-aside areas motivated by vertebrate protection needs may also be invaluable sanctuaries for native insects largely extirpated from the wider enveloping landscape. But the specific management needed for focal vertebrates cannot 'automatically' cover the need of all those insects. In most such situations, inventories of the insects present have not been made as foundation for more refined management—other than, in some cases, noting individual 'listed' species signalled for conservation concern. Such listing is an important part of the political dilemma in creating commitment to conservation investigation and action and as a formal basis for lobbying for support.

3. The scientific dilemma, that basic scientific knowledge of insects is scarce and underfunded. Linking with the four scientific shortfalls, above, all of which need vastly increased fundamental information on taxonomy and/or ecology, the low level of societal interest in such pursuits is an accelerating concern as the scale of need becomes increasingly apparent.

   The decline of traditional taxonomy is a major impediment to documenting the world's insects. The importance of species definition and enumeration underpins much description of biodiversity, but the number of classical insect taxonomists continues to decline. As specialists retire, they are rarely replaced by people with equivalent knowledge or parallel interests, and succession planning to assure continuity of expertise in major collections and museums is highly unusual. In part, this reflects changes in institutional funding priorities but also changes in university curricula, so that even training in basic entomology has become very restricted, but also the wider declines in interests in 'natural history' as the previously widespread recruiting ground for much perception of insect diversity. Simply, 'describing and counting species' is viewed widely as an unrewarding career path, and relevant education and training has declined to be replaced by more technical and 'modern' disciplines seen as more likely to lead to employment. Many of the remaining insect taxonomists in museums are based in regions in which the local insect faunas are well-known, and many cannot address at first hand the pressing problems of assessing insects in other parts of the world where needs for conventional descriptive taxonomic work are greatest.

Likewise, basic inventory studies of insects and the sampling programmes needed to establish patterns of distribution and abundance are increasingly difficult to fund and sustain. Seeking statistical and surrogate shortcuts to further reduce need for fundamental biological studies through extrapolations and 'modelling' from limited data lead to further neglect of basic more comprehensive documentation. Developments of suites of taxonomic procedures to describe species at a faster rate, notably by using DNA barcoding and metabarcoding, are receiving much attention, and the more general skills needed to use such methods are attractive because they are 'transferable' across different taxa so that, rather than electing to become a specialist in a particular family of moths, flies or beetles, with patently limited employment opportunities, a molecular biologist capable of applying sophisticated differentiation techniques may have a far wider career path to tread. As Leather (2009) and others have emphasised, the decline of entomology (indeed, invertebrate) taxonomy training in universities means that, even in the United Kingdom as a traditional stronghold for insect natural history, retiring specialists cannot be replaced simply because potential successors have not been encouraged or allowed to develop their interests formally. Leather described the biology departments of United Kingdom universities as 'bereft of entomologists'. In Australia, as elsewhere, many diverse, significant and under-recorded groups of insects have no resident specialist by whom specimens can be identified.

If the widespread maxim that it is desirable to name all species before effective conservation is followed, it is likely that far more species will be lost before this can be achieved—if it is ever possible. Since Wilson (2003) initiated the concept of an 'Encyclopedia of Life' with the proposition that it might be possible to discover, describe and name all the world's species within a few decades, the logistics involved have been discussed extensively. Considerable doubts have been advanced; May (2004), for example, concluded it unrealistic to complete this task within a generation unless the total number of species is far fewer than generally supposed. One key hindrance is belief that suggested 'rapid aids' such as DNA analyses and web-based publications do not replace conventional taxonomic expertise. As Scotland et al. (2003) implied, 'added technology does not necessarily increase taxonomic knowledge, as opposed to enhanced investment in taxonomic expertise'. Educated discussion (Gonzalez-Oreja 2008) emphasises, repeatedly, the paucity of taxonomists. For any such complete 'Encyclopedia', for which insects are a substantial component, and depending on a designated adequate rate of descriptions (X species/taxonomist/year), he suggested that some 236,000 taxonomists were needed at that time and every year for the next 25 years. This prospect seems unrealistic when compared with a then current estimate of 6000–7000 active taxonomists and the scarcity of training, equipment and financial opportunity, as well as perceived career opportunities. The alternative ambition suggested by Gonzalez-Oreja is a highly abbreviated 'Brochure of Life' with a relatively small proportion of species described, and to which conservation priorities will be largely restricted as the only taxa that are sufficiently noticed.

'Naming species', however desirable in providing a firm basis for authoritative inventory and diversity assessments, does not prevent extinction. Gonzalez-Oreja (2008) affirmed that the taxonomic impediment was then (1) obstructing advancements in sound management of known species, and (2) obstructing the discovery of unknown ones, through failure to recognise these—in short, the paucity of taxonomic knowledge and of experts able to remedy this limit 'our ability to conserve, use and share the benefits of biological diversity'. Those restrictions underpin the importance of the Linnean shortfall; Eggleton's (2020) contention that no more than 20% of insect species thought to exist have been described endorses his inference that 'we cannot easily identify which species it is important to conserve'.

However, and as Janzen (1994, 1997) and others have emphasised, completing an inventory, even when valuable in appraising biodiversity, is a very different exercise from conservation. In contrast to sentiments such as 'we cannot conserve it unless we know what is there', Janzen considered that the substantial delays in achieving this state as a basis for conservation does not accept the urgency of conservation need and that many years may be lost over the delay period needed. Janzen's comments continue to resonate strongly in relation to the purposes of labour-intensive inventory surveys of tropical insects, and were framed as 'what do we not need to know?'

His major points (discussed further by New 2007) included (1) that we do not need to enumerate all the taxa present because we already known that there are a lot (thousands to millions) and that most are wholly or almost wholly unknown; (2) detailed distributional knowledge of all species is not needed, because the areas available for conservation are already mostly defined, and the fauna of those areas is a paramount focus for study; (3) detailed data on individual species is largely redundant in the wider biological and sociological contexts of conservation need; and (4) realistic triage is needed to harmonise conservation with agroecology in the tropics, perhaps with wider attention to larger areas and (possibly) less to small isolated areas. The latter are a consistent priority for much temperate region species-focused conservation, but Janzen's pragmatic statement that 'the living dead and the population fragments sprinkled across the tropical agroscape are slated for the dust bin', however fatalistic it appears, is also realistic. Unlike the attention to saving individual Endangered or other Threatened insect species pioneered in the northern temperate regions, that approach becomes manifestly inadequate—perhaps even unwise in diverting limited support from wider needs—in the tropics. Insects in the tropics and much of the southern temperate areas, potentially benefit from a broader perspective involving emphasis on community or 'habitat' conservation in those environments in which numerous under-documented insects are suspected to have similar levels of concerning conservation need, but will never be proven to do so.

2.4 Recognition of Threatened Insect Species

One significant context is need to identify reliably the species that have been 'listed' or nominated for conservation need and priority, having been evaluated in some way as threatened, and through whose study much of insect conservation has been driven. Those species can become important 'flagships' for conservation awareness. That formal recognition often accords those species priority for the limited funding and other support available—and thereby reduces the resources available for other species. 'Listing species' under any conservation legislation or recommendation is therefore a responsible action, and evidence of need must be expressed as thoroughly and lucidly as possible. That evidence includes specifying the identity of the target species, its distribution and population features, its life history and ecology (as far as can be stated, and most notably the critical resources needed), the threats (historical, current and potential future threats) to the species and possible remedial management. In Australia, nominations at state/territory or commonwealth level (made by any concerned person or organisation) are assessed by panels, who may need (or be obliged) to seek independent peer review, so that other general information and key references to support a nomination of a poorly known insect are useful, but formal requirements for a nomination differ across jurisdictions. 'Listing' implies future protection, but often conservation does not proceed beyond this simple notification of need. Lists of protected or threatened insect species are almost invariably strongly biased in taxonomic coverage. Most are dominated by butterflies, reflecting their strong public appeal and wider concerns for their fate accompanied by credible evidence for losses and declines in abundance from (in Europe) a century and more of incidence records. Odonata, Orthoptera and Coleoptera are also reasonably represented—but only seven of the 25 insect orders in Europe are on any protected species lists (Leandro et al. 2017). Two of the largest orders (Hymenoptera, Diptera, together comprising more than 40% of European insect species) were then absent. Leandro et al. noted that only 123 of the 105,000 known European insect species were listed, clearly a severe understatement of real need and not taxonomically representative; but also capitalising on knowledge from popular flagship groups from which increased interest can grow or be fostered. Parallel trends are common elsewhere, with little formal acknowledgement of conservation needs of most insect groups. Calls to increase species numbers and taxonomic representation, even when heeded, are unlikely to ever render them fully representative of rich and diverse insect faunas—and the wider dilemma is that whilst lists of threatened insects are too short to be representative, they are already also too long for all listed candidates to be fully supported. Although formal 'listing' may oblige, or facilitate, preparation of some further assessment in the form of an 'Action Plan' or 'Action Statement' as a prelude to management, this often remains unaccomplished or considerably outdated. However, as Leandro et al. (2017) put it 'protected species will nonetheless still represent a very small proportion of the huge diversity of entomofauna'. The limitations of focusing solely on individual species in conservation are reinforced by such statements. Many little-known or difficult to identify species cannot be recognised by other than a few specialists and cannot become realistic candidates for credible listing in this way; however well-intentioned and deserving they may be, listing such taxa can only cause confusions. Likewise, credibility can also suffer from the well-intentioned or precautionary listing of numerous species to represent local faunas and which include common and non-threatened taxa. The numerous species of butterflies listed under the Indian Wildlife Protection Act 1972 are one such example.

Table 2.2 Four categories of narrow-range endemism recognised for Australian butterflies, each including taxa previously assessed as having conservation concern (from New and Sands 2002)

1. Only one or extremely few (<5) populations over a very small range (<50 km)
2. Very few to several (<40) populations over a small range (<200 km)
3. Very few (<10) fragmented populations over a wider range (>300 km)
4. Possible/actual hybrid zones with narrow ranges

Evaluation of increasing numbers of insects for specific threat continues to display the varied and increasing needs of many, with potential for repeated appraisals at intervals and using the same criteria to detect temporal trends and to reflect both richness and vulnerability of local faunas and different taxa. 'Red-Listing' species on the extent of threats they face and their distributional trends is an important tool in promoting national and international conservation interest, in providing objective information on the better-known taxa and showing gaps in knowledge for many others. Not in themselves legislative, Red Lists have sometimes been transformed to become the basis of more formal legislation. However, legislative scope and obligations can differ substantially in completeness and outcomes, as evident from differences across different European countries (Collins 1987 and many later accounts, such as Van Swaay and Warren 1999). Many European insects are more restricted endemics with highly restricted ranges on the continent. Many others, in contrast, are distributed patchily across the continent, so that conservation need may apply only to particular countries or regions.

'Endemism' is assessed most commonly as restriction to a stated country or region, so that in Europe many insects occur only in particular countries, or in circumscribed regions such as the alpine or southern (Mediterranean) areas. Many species are termed 'narrow-range endemics' and occur only on very small ranges—some from only single sites or very restricted biotopes. Extensive study of much of the European fauna renders such evaluations credible but in many other places simply that a recognisable species has been recorded only so sparsely may reflect dearth of collecting and survey over larger areas, rather than confirmed absence. In some instances, perhaps possible only from the stimulus from 'listing', subsequent surveys have recorded the species elsewhere and so distributed more widely than earlier supposed; in other cases, additional searches have tended to confirm the earlier impression of scarcity. Putative localised distributions can mirror several rather different situations, such as (1) a natural population that has not been influenced in any significant way by human activity, as narrow-range endemicity; (2) a pattern that reflects the outcome of habitat loss and fragmentation, either as a population with a presumed wider historical range or (more rarely) as evolutionary differentiation or speciation within discrete habitat patches; and (3) a pattern that reflects contractions at the extremes of a former range. Four categories of narrow-range endemicity were distinguished for Australian butterflies (New and Sands 2002), each with a variety of examples (Table 2.2). Interpretation of those patterns can dictate the need and extent of conservation management needed, recognising

Table 2.3 Overview of the IUCN Red List criteria of threat. The three categories of threat are critically endangered (CR), endangered (E) and vulnerable (V). (AOO, area of occupancy; EOO, extent of occurrence) (IUCN 2017)

A. Population reduction (over 10 years or three generations)
CR: (>90%, AND causes are reversible, understood and ceased) OR > 80%
E: (>70% AND causes are reversible, understood and ceased) OR > 50%
V: (>50% AND causes are reversible, understood and, ceased) OR > 30%
B. Geographic range
CR: (EOO < 100 km^2 OR AOO <10 km^2 AND two of: (a) fragmentation and/or a single location; (b) continuing decline; (c) extreme fluctuations.
E: (EOO <5000 km^2 OR AOO <500 km^2 AND two of: (a) fragmentation and/or locations <5; (b) continuing decline; (c) extreme fluctuations.
V: (EOO <20,000 km^2 OR AOO <2000 km^2 AND two of: (a) fragmentation and/or locations <10; (b) continuing decline; (c) extreme fluctuations.
C. Small population size and decline
CR: n < 250 AND (reduction >25% over 3 years or one generation OR (reduction AND (larger subpopulation <50 OR > 90% individuals in a single subpopulation OR extreme fluctuations)))
E: n < 2500 AND (reduction >20% over 5 years or two generations OR (reduction AND (larger subpopulation <250 OR > 95% individuals in a single subpopulation OR extreme fluctuations)))
V: n < 10,000 AND (reduction >10% over 10 years or three generations OR reduction AND (larger subpopulation <1000 OR 100% individuals in a single subpopulation OR extreme fluctuations)))
D. Very small or restricted population
CR: n < 50
E: n < 250
V: n < 1000 OR AOO < 20 km^2 OR locations <5
E. Quantitative analysis of extinction risk
CR: >50% over 10 years or three generations
E: >20% over 20 years or five generations
V: >10% over 100 years

that the array of insect species on any site, or in any biotope, will include those that transcend the major categories recognised for British butterflies as 'habitat specialists' and 'wider countryside specialists' (Pollard and Eversham 1995). The former are confined to discrete habitat patches, have largely declined in modern landscapes, tend not to disperse far or use linear habitats such as roadsides, and are usually univoltine and monophagous. The latter have a broader habitat range and are commonly more mobile as well as being bivoltine or multivoltine, and often polyphagous. These principles are likely to have much wider relevance to assessing insect needs.

The distribution of an insect species is a core element of defining its conservation status. The IUCN (World Conservation Union) Red List criteria (Table 2.3 and Fig. 2.2) include both 'extent of occurrence' and 'area of occupancy', the former a mapping extension to encompass the sites from where the species is known, so

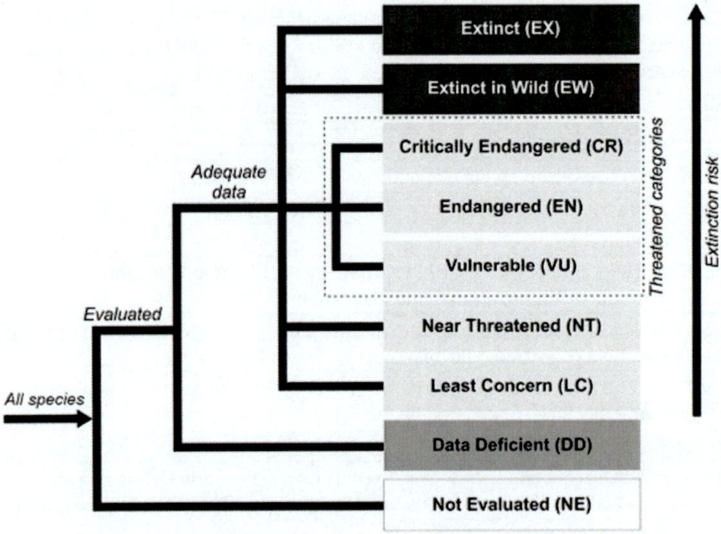

Fig. 2.2 The hierarchy of IUCN Red List categories: the groupings to which assessed species may be allocated, with risk of extinction for evaluated taxa increasing toward top of page. Based on figure from IUCN (2012, with permission)

embracing the most likely areas where additional populations might occur with targeted surveys, and the range of metapopulations, among which some suitable sites might not be occupied all the time. This important principle is not appreciated sufficiently in much insect conservation planning, in which the most frequent presumption is of a (nominally) more simple closed population structure, so that individual site focus is critical, and the wider landscape context of the species not properly acknowledged. Uncertainty again emphasises the wisdom of undertaking long-term studies wherever possible, not least to infer whether disappearance from a site is a real concern or part of a wider metapopulation pattern of natural rolling colonisations and extirpations involving suitable sites across the landscape. The important difference between 'small populations' and 'declining populations' (in the sense of Caughley 1994) is that the former may be entirely natural and stable—although of course not immune from any future threats that arise. Simply because an insect species is known only from a single locality or small range, as a putative narrow-range endemic, does not invariably imply that it should 'automatically' be given high conservation status—although it may strongly influence any such decision. Two practical contexts emerge by correlating restricted range with rarity and/or vulnerability (Australian butterflies: New and Sands 2002). First, for many ecologically specialised species, much potential and apparently suitable habitat is not occupied for usually unknown reasons. This situation endorses that it may not be valid to extrapolate an insect's incidence to the extent of habitat available, even if conditions such as climate and topography are similar across occupied and unoccupied sites. Second, the real definition of a species' range and the definition of breeding habitat is also difficult in regions that are largely unexplored: for some species, the few sites from where they have been recorded have not been proved to be breeding sites. Some topographical features (such as hilltops on which butterflies and other insects may gather from a wide surrounding area and facilitate mate-finding) exemplify this, but bias is emphasised by collectors continuing to visit such 'traditional' localities to seek specimens of otherwise elusive species based on known history. Exploration of the wider enveloping landscape is thereby often neglected, but with some presumption that the target species indeed breeds 'somewhere not too far away'.

In Red-Listing, each species is evaluated against the criteria for allocation to one of eight categories (IUCN 2017) to reflect whether it is threatened and, if so, how severely. A 'threat grade' under any one of the five criteria is sufficient to treat the species as threatened, and the threat level increases as the numbers of criteria give similar outcomes. This system is regarded as a global standard for ranking threats, as explicitly including threatening processes in evaluation rather than simply the conditions of small populations or restricted distributions, both of which can be entirely natural, and are widespread among insects. The IUCN criteria consider incidence and extent of reductions of population size and geographical range. However, and reflecting the impediments discussed earlier and by Cardoso et al. (2011a, b), rigorous data to estimate threat status of most insects is difficult to obtain. The practical problems include the naturally high population fluctuations in many taxa, even over consecutive generations or other short time spans and influenced by sensitivity to environmental factors such as weather, and the difficulties of adequate standardised sampling for species with poorly known biology. Applications of IUCN criteria A, C, D and E are commonly very tentative—perhaps especially for 'E' when population viability exercises for insects are largely impracticable. Even for butterflies and larger moth species in Britain, for which reliable long-term monitoring data are available, population viability confounded assessment of extinction risk using the Red-List criteria (Fox et al. 2019), especially the time-specified criterion A (population reduction over 10 years) and for which a longer period might increase reliability in overcoming the short-term vagaries in numbers that lead to considerable inter-year differences in many documented insect populations—and, by implication, numerous others throughout the world. The purpose of the '10-year rule' is to show recent trends of loss in risk assessments while not heeding earlier historical declines of species that are now stable or recovering and so not threatened. Examples noted by Fox et al. suggested that longer time series are needed to provide sufficient information for more robust assessments. Such extended surveys are almost always unavailable or impracticable, but for European butterflies has allowed development of highly credible Red Lists and designation of priority areas for conservation (Van Swaay and Warren 2006).

Debate over the relevance and suitability of the IUCN criteria for invertebrates focuses on the general lack of information and the difficulties of interpreting vulnerability by any quantitative means, as well as differences in reliability across different taxonomic groups. Thus, Cardoso et al. (2011b), in listing shortcomings of

Table 2.4 Shortcomings of the current IUCN criteria of threat (Table 2.3) listed by Cardoso et al. (2011b) (AOO, area of occupancy; EOO, extent of occurrence)

Criterion	Shortcomings
A. Population reduction	Abundance decline difficult to measure
B. Geographic range	EOO and AOO very large for invertebrates. Hard to meet all requirements (small range, fragmentation, decline), incomplete sampling
C. Small population size and decline	Inadequate threshold values for abundance
D. Very small or restricted population	Inadequate threshold values for abundance
	Abundance data hard to obtain
E. Quantitative analysis of extinction risk	Very difficult to apply without a large amount of data
F. Co-extinction	Not considered

the criteria (Table 2.4), suggested more extensive reliance on distributional criteria and also the inclusion of co-extinctions as a component. The latter was discussed more extensively by Moir and Brennan (2020) as a subcomponent of criterion D that involved defining conservation of dependent species based on relationship to their host(s) as 'co-threatened taxa'.

One particular point of concern has been the application of the '10-year rule' for Criterion A, as discussed by Fox et al. (2019). More than 40 years of continuous records for British butterflies and larger moths provide two unique data sets for analysis. Fox et al. used these to examine the effects of varying the commencement year of a 10-year recording period by a single year and repeating this with the same data to provide five different decadal records each changed by only a year. They found that even a single year change had considerable effects on inferring whether a species was formally threatened: thus for the 109 moth species that qualified as threatened, only five did so in all five time sequences examined, and the remaining 104 species were threatened under only some of those decades. For Criterion 2 (reduction in population size) one of those differences reduced the number of threatened species from 62 to 20.

Even for such long data sets—far more than any available in Australia—reliable appraisal of threat status is difficult. Fox et al. pointed out the two major sources of error in any such time series as (1) imperfect sampling, so 'observational errors', and (2) short term stochastic variation in population levels caused by environmental variability, as 'process errors', so that inaccurate detection of population trends can give either false positives (incorrectly classifying a species as threatened) or false negatives (failing to classify a deserving species as threatened). In general, estimates of threat status must necessarily be based on far shorter observation periods.

However, attempts to render the IUCN criteria more useful for invertebrates have included increasing the current 10-year period to 25 years, as for European butterflies (van Swaay and Warren 1999), but these authors also dismissed use of criteria C, D and E as irrelevant or inapplicable, even for this best-known regional insect fauna. For Australia, Clarke and Spier-Ashcroft (2002) noted possible value of

Table 2.5 Scheme for assessing conservation status of Australian butterflies, using IUCN threat category terms modified to apply to Australian butterflies in largely avoiding quantitative thresholds (from Sands and New 2002)

1. Information on biology, distribution and resident/vagrant status sufficient to make an informed evaluation of conservation status	2
– Information insufficient to make an informed evaluation: with little or no information on any of the above topics	Data deficient
2. Threats defined for the species: threats to major populations or population segregates likely to lead to species extinction	3
– No threats defined for the species: threats to major populations or population segregates not likely to lead to species extinction	no conservation significance
3. Threats identified for all known populations and considered to pose a risk of extinction within 5 years, usually no more than five populations or major population segregates known	Critically endangered[a]
– Threats identified for all or most known populations, normally including those of greatest significance (size, distribution), and considered to pose a risk of extinction within 5–10 years	Endangered[a]
– Threats identified for some populations and considered sufficiently important to pose a risk of species extinction within 10–20 years	Vulnerable[a]
– Threats identified but not considered to pose a risk of extinction within 20 years	Lower risk[a]

[a]For each category of threat listed conditions, specified by Sands and New, are implicit

ranking on less precise data (as suggested by Akcakaya and Ferson 1999; Akcakaya et al. 2000)—but for the taxa they trialed the approach gave polarised outcomes whereby a species was either Critically Endangered or Data Deficient. In short, the IUCN criteria, however valuable as a guide, are difficult to rely on in providing definitive rankings of threat of extinction for most Australian insects, and there is need for less quantitative alternatives that can build on any useful information available but without insisting on rigid quantitative data on population changes. Modified schemes for assessing conservation status of butterflies in Australia (Sands and New 2002, Table 2.5) and New Zealand (Patrick and Dugdale 2000, Table 2.6) exemplify such attempts. Thus, Sands and New recognised the IUCN categories but modified the criteria for allocating species to them by minimising the quantitative information needed. Whatever scheme is used, the primary need is for tangible detection of threats and some evaluation of their severity and impacts, but also with sufficient flexibility in practical application to cater for the peculiarities and idiosyncrasies of individual species and their environments.

Perhaps more frequently in Australia, evaluation of threat arises from notice of impending or actual loss of the site(s) on which a localised or otherwise significant insect species is known to occur, and the stimulation this provides for urgent investigation, essentially as 'crisis management'. In Victoria, the endemic flightless stonefly *Riekoperla darlingtoni* is known only from the summit of a single mountain (Mount Donna Buang), where it is associated with a small stream. Construction of a car park to service tourism, including sightseeing in winter, led to concerns over

Table 2.6 Attributes used for New Zealand Lepidoptera to give species 'conservation significance'. Each species was allocated to a single category, deemed the 'most significant' for that species (Patrick and Dugdale 2000). See also Table 2.7 for the broader approach later adopted in New Zealand

A. Uncommonly or rarely encountered but widespread and with no historical evidence of dwindling populations or range contraction
B. As above but with historical evidence of dwindling populations and range contraction
C. Known only from the type locality (incudes species for which the type is lost)
D. Uncommon or rare, biology unknown
E. Type locality grossly altered
F. Type locality at risk
G. Host plant/niche at risk, or predator influence seen, in major part of species range
H. Genetic swamping of the endemic population by self-adventive sister taxon
I. No record of capture for >25 years, and now possibly extinct

stream pollution from runoff or possible spillage exterminating the stonefly, and listing it as threatened—but with rather different appraisals under the Commonwealth Environment Protection and Biodiversity Conservation Act (EPBC Act) and Victorian Flora and Fauna Guarantee Act (FFG Act) creating some confusion (New 2008). In another classic case in Victoria, the Eltham copper butterfly (*Paralucia pyrodiscus lucida*, Lycaenidae) was widely believed to have been exterminated from its discovery areas north east of Melbourne by urban expansion, when a population was discovered in 1987 on a site scheduled for imminent housing development (p. 202). That discovery led to an extensive conservation campaign, rapid listing of the subspecies under FFG, and protection of the discovery site as one of Australia's first dedicated butterfly reserves. Other parallel cases occur, in which unexpected discovery has created a sense of urgency that, supported by sound advocacy, has engendered recognition of the insect being 'threatened', with the impacts of that threat not defined formally, but in some cases not as obviously severe as complete destruction of all known occupied habitat as in the above examples.

Reflecting difficulties and ambiguities of defining that severity, Sands and New (2002, 2003) also considered that 'listing' an insect species should be linked also with the facility to 'delist', to avoid regulatory or advisory lists becoming misleading if non-threatened species are retained. This intention covered both (1) species listed unjustifiably and (2) those that have subsequently become assessed as non-threatened through either (a) work following initial listing revealing greater security—for example, by discovery on additional sites or of secure populations, or (b) practical conservation management has led to recovery. For the last, Sands and New suggested the term 'rehabiitated species' that, after delisting should remain on a schedule for periodic assessment in order to ensure that they are not simply abandoned, that threats do not recur and again lead to concern, and so that the (perhaps substantial) conservation investment is protected. The sentiment is echoed in the 'recovery' category of risk used in New Zealand (see Table 2.7).

The aim of the IUCN criteria—to provide a system applicable to all species and to avoid idiosyncratic and less clearly understood modifications for particular taxonomic groups—has been adopted widely, although modifications will surely continue to be suggested. Limitations of applications to insects, and the levels of uncertainty that arise, will continue to frustrate assessors and arise largely from lack of survey for more precise definition of geographical ranges, and limited taxonomic capability. Many have been discussed for Australian contexts (Hutchings and Ponder 1999) and in that workshop a number of themes were discussed: not least, the frequent uncertainty over the structure of a 'population' to which any criteria are applied. As noted at the start of this chapter, the vulnerability of a true 'closed population' on a site may be far different from that of a metapopulation segregate, for example. Insects with a metapopulation structure will normally undergo local extirpations from individual sites within a larger range, and those sites may be re-colonised at some later stage, providing that landscape changes have not led to loss of connectivity.

An approach taken for evaluating status of New Zealand Lepidoptera, following the earlier appraisal by Patrick and Dugdale (2000), categorised taxa against the 'New Zealand Threat Classification System' (NZTCS) devised by Townsend et al. (2008). The points of NZTCS are summarised in Table 2.7, for comparison with the IUCN criteria, and the 202 taxa appraised by Hoare et al. (2017) included 70 that were classed as 'taxonomically indeterminate', as taxa accepted as distinct for assessment but that did not have valid published names. That step could usefully be taken effectively in Australia to help accommodate the numerous recognised but undescribed insect taxa. Although an undescribed ant is listed under Victoria's legislation, this—and broader assessment of undescribed taxa—is unusual. For completeness, NZTCS also listed the categories of 'Non-resident Native' taxa, 'Not Threatened' taxa and 'Introduced and Naturalised' taxa.

Collen and Bohm (2012) considered that a concerted effort to systematically add species from under-represented groups to the global Red List would have far greater impact on coverage of invertebrates and may contribute more understanding to the risks of extinction. Nevertheless, the paucity of information may well lead simply to increased numbers of species for which the only realistic status allocation is 'Data Deficient' (DD), in itself useful in implying needs for future efforts.

One limitation of trying to assess even the best-documented insect groups outside the northern temperate region is simply their richness combined with inability to diagnose them properly, so that it is still impossible to assess all the species, many of them very poorly known. For butterflies, Lewis and Senior (2011) promoted the approach of a 'Sampled Red List Index' (SRLI) by which a taxonomically and geographically representative subset is used. Earlier study (Baillie et al. 2008) suggested that an index sample of 1500 species was sufficient, even when many are DD. Lewis and Senior noted that existing Red List butterfly assessments were 'patchy and idiosyncratic', also with subjective or out-of-date categories and criteria, as well as very uneven taxonomic and geographical coverage. In order to remedy this for global representation, the projected composition by geographical region of species to be sampled (Table 2.8) was based on estimated regional species richness, within which each geographical 'pool' species selection was random. However, a proportion (estimated as 15%, and likely to be far higher for most other insect

Table 2.7 Categories and criteria for species conservation status assessment under the New Zealand Threat Classification Scheme (Molloy et al. 2002; Townsend et al. 2008)

Data deficient
Taxa suspected to be threatened or, in some cases, possibly extinct, but are not definitely known to belong to any particular category due to lack of current information
Threatened
Taxa that meet the criteria for Nationally Critical, Nationally Endangered and Nationally Vulnerable
Nationally critical:
A—Very small population (natural or unnatural)
A(1) <250 mature individuals, regardless of cause
A(2) 2 or fewer subpopulations, <200 mature individuals in the larger subpopulation
A(3) Total area of occupancy <1 ha (0.01 km^2)
B—Small population (natural or unnatural) with a high ongoing or predicted decline
B(1/1) 250–1000 mature individuals, predicted decline 50–70%
B(2/1) <5 subpopulations, <300 mature individuals in the largest subpopulation, predicted decline 50–70%
B(3/1) Total area of occupancy <10 ha (0.1 km^2), predicted decline 50–70%
C—Population (irrespective of size or number of subpopulations) with a very high ongoing or predicted decline (>70%)
Nationally endangered:
A—Small population (natural or unnatural) that has a low to high ongoing or predicted decline
A(1/1) 250–1000 mature individuals, predicted to decline 10–50%
A(2/1) <5 subpopulations, <300 mature individuals in the largest subpopulation, predicted to decline 10–50%
A(3/1) Total area of occupancy <10 ha (0.1 km^2), predicted decline to 50%
B—Small stable population (unnatural)
B(1/1) 250–1000 mature individuals, stable population
B(2/1) <5 subpopulations, >300 mature individuals in the largest subpopulation, stable population
B(3/1) Total area of occupancy <10 ha (0.1 km^2), stable population
C—Moderate population and high ongoing or predicted decline
C(1/1) 1000–5000 mature individuals, predicted decline 50–70%
C(2/1) <15 subpopulations, <500 mature individuals in the largest subpopulation, predicted decline 50–70%
C(3/1) Total area of occupancy <100 ha (1 km^2), predicted decline 50–70%
Nationally vulnerable:
A—Small, increasing population (unnatural)
A(1/1) 250–1000 mature individuals, predicted increase >10%
A(2/1) <5 subpopulations, <300 mature individuals in the largest subpopulation, predicted increase >10%
A(3/1) Total area of occupancy <10 ha (0.1 km^2), predicted increase >10%
B—Moderate, stable population (unnatural)
B(1/1) 1000–5000 mature individuals, stable population
B(2/1) <15 subpopulations, <500 mature individuals in the largest subpopulation, stable population

(continued)

Table 2.7 (continued)

B(3/1) Total area of occupancy <100 ha (1 km^2), stable population
C—Moderate population, with population trend that is declining
C(1/1) 1000–5000 mature individuals, predicted decline 10–50%
C(2/1) <15 subpopulations, <500 mature individuals in the largest subpopulation, predicted decline 10–50%
C(3/1) Total area of occupancy <100 ha (1 km^2), predicted decline 10–50%
D—Moderate to large population and moderate to high ongoing or predicted decline
D(1/1) 5000–20,000 mature individuals, predicted decline 10–50%
D(2/1) <15 subpopulations and <1000 mature individuals in the largest subpopulation, predicted decline 30–70%
D(3/1) Total area of occupancy <1000 ha (10 km^2), predicted decline 30–70%
E—Large population and high ongoing or predicted decline
E(1/1) 20000–100,000 mature individuals, predicted decline 50–70%
E(2/1) Total area of occupancy <10,000 Ka (100 km^2), predicted decline 50–70%
At risk
Taxa that meet the criteria for declining, recovering, relict and naturally uncommon
Declining:
A—Moderate to large population and low ongoing or predicted decline
A(1/1) 5000–20,000 mature individuals, predicted decline 10–30%
A(2/1) Total area of occupancy <1000 ha (10 km^2), predicted decline 10–30%
B—Large population and low to moderate ongoing or predicted decline
B(1/1) 20000–100,000 mature individuals, predicted decline 10–50%
B(2/1) Total area of occupancy <10,000 ha (100 km^2), predicted decline 10–50%
C—Very large population and low to high ongoing or predicted decline
C(1/1) >100,000 mature individuals, predicted decline 10–70%
C(2/1) Total area of occupancy >10,000 ha (100 km^2), predicted decline 10–70%
Recovering:
Taxa that have undergone a documented decline within the last 1000 years and now have an ongoing or predicted increase of >10% in the total population or area of occupancy taken over the next 10 years or three generations, whichever is the longer
A—1000–5000 mature individuals or total area of occupancy <100 mha (1 km^2), and predicted increase >10%
B—5000–20,000 mature individuals or total area of occupancy <1000 ha (10 km^2), and predicted to increase >10%
Relict:
Taxa that have undergone a documented decline within the last 1000 years and now occupy <10% of their former range and meet one of the following criteria:
A—5000–20,000 mature individuals: Population stable (±10%)
B—>20,000 mature individuals, population stable or increasing at >10%
Naturally uncommon:
Taxa whose distribution is confined to a specific geographical area or which occur within naturally small and widely scattered populations, where this distribution is not the result of human disturbance

Table 2.8 Representation of butterflies by biogeographic realms proposed for assessment under the IUCN Sampled Red List Index (Lewis and Senior 2011)

Biogeographic realm	Estimated % of global fauna	No species to be assessed
Neotropical	42	630
Afrotropical	21	315
Oriental	14	210
Australasian	8	120
Palaearctic	11	165
Nearctic	4	60
Total	100	1500

groups) were categorised as DD and for which extinction risk could not be evaluated. The status of Data Deficient insects generally extends across all other threat categories because it can apply to taxa that are rarely seen, so inferred to be threatened or, even, extinct.

For Odonata, a randomly selected sample of 1500 species—more than a quarter of the global total of described species—implied that around 10% of species were threatened with extinction (Clausnitzer et al. 2009), but more than half were of Least Concern. Clear geographical trends occurred, with lowest proportion threatened in the Nearctic and the highest proportion in the tropical regions, notably the Indo-Malayan region. Clausnitzer et al. found the relatively low levels of threat to Odonata unexpected (and among the lowest such group estimates found), and attributed this in part to a combination of wide habitat range (including degraded waterbodies), the strong colonisation potential of many taxa through active flight, and facility to use small habitat patches. Most Odonata, in contrast to some other showy insect groups, are largely unaffected by hunting or trade. However, Clausnitzer et al. also noted 'extensive Data Deficiency' among taxa of some key areas.

A related concern, even in well-documented faunas, is the practice of 'precautionary listing' of insect species—that is, for cases of doubt over a species' real conservation need, listing it to assure notice as a possibly threatened species, with implied conditions of Data Deficient. Operational opinion can become polarised, as (1) 'it pays to be careful, in case . . .', or (2) 'to be credible, listing should consider only the genuinely threatened species'. Especially in poorly-known faunas, the former approach could lead rapidly to extremely long lists of poorly documented taxa known from few individuals or sites and unlikely to be studied further. The latter approach is also likely to be highly incomplete, in simply representing the taxa (however valid and accurately assessed) that have come to notice. Neither approach may closely approximate the real scale of need.

Two conditions of listing an insect, suggested for butterflies by Kudrna (1986) but often overlooked, have major practical value. Rather than listing being simply a notice with some potential to lead to further protection or conservation action, Kudrna considered that listing species was worthwhile only if (1) 'It must protect them and their habitat from the effects of harmful anthropogenic factors primarily responsible for their decline', and (2) 'It must facilitate the continuing of genuine research, whether it is carried out by professional or voluntary research workers'.

Moving from species considerations to documenting wider biodiversity, either by species richness or by variety of ecological functions and ecological residence, constraints occur at both 'descriptive' (characterisation of assemblages or communities) and 'comparative' (across sites and time) levels, in which taxon recognition and consistency of that recognition may have far-reaching consequences if interpretations are inaccurate and/or inconsistent. Conservation policy and practice can then rapidly lose credibility. Both alpha-diversity (the total number of species present in an area) and beta-diversity (the compositional differences between assemblages in different areas) are relevant and together constitute gamma-diversity (the number of species in a larger area, and for which detection is often the main purpose of inventory studies). A general point across many insect surveys purporting to display diversity, perhaps especially in remote areas and much of the tropics, is that sampling intensity is very limited, and the insects often not identified to species level (p. 59), so that comparisons across sites or treatments may not accurately define 'real diversity', as sometimes claimed.

References

Akcakaya HR, Ferson S (1999) RAMAS red list: threatened species classification under uncertainty. User manual, Setauket, New York

Akcakaya HR, Ferson S, Burgman MA, Keith DA, Mace GM, Todd CR (2000) Making consistent IUCN classifications under uncertainty. Conserv Biol 14:1001–1013

Baillie JEM, Collen R, Amin R, Akcakaya HR, Butchart SHM et al (2008) Toward monitoring global biodiversity. Conserv Lett 1:18–26

Beard JJ, Ochoa R, Braswell WE, Bauchan GR (2015) *Brevipalpus phoenicis* (Geijskes) species complex (Acari: Tenuipalpidae) – a closer look. Zootaxa 3944:1–67

Cardoso P, Erwin TL, Borges PAV, New TR (2011a) The seven impediments in invertebrate conservation and how to overcome them. Biol Conserv 144:2647–2655

Cardoso P, Borges PAV, Triantis KA, Ferrandez MA, Martin JL (2011b) Adapting the IUCN red list criteria for invertebrates. Biol Conserv 144:2432–2440

Caughley G (1994) Directions in conservation biology. J Anim Ecol 63:215–244

Clarke GM, Spier-Ashcroft F (2002) A review of the conservation status of selected Australian non-marine invertebrates. Environment Australia, Canberra

Clausnitzer V, Kalkman VJ, Ram M, Collen B, Baillie JEM et al (2009) Odonata enter the biodiversity crisis debate: the first global assessment of an insect group. Biol Conserv. https://doi.org/10.1016/j.biocon.2009.03.028

Collen B, Bohm M (2012) The growing availability of invertebrate extinction risk assessments – a response to Cardoso et al. (October 2011): adapting the IUCN red list criteria for invertebrates. Biol Conserv 149:145–146

Collins NM (1987) Legislation to conserve insects in Europe. Pamphlet no 13, Amateur Entomologists' Society, London

Eggleton P (2020) The state of the world's insects. Annu Rev Environ Res 45:8.1–8.22

Fox R, Harrower CA, Bell JR, Shortall CR, Middlebrook I, Wilson RJ (2019) Insect population trends and the IUCN red list process. J Insect Conserv 23:269–278

Gonzalez-Oreja JA (2008) The encyclopedia of life vs. the brochure of life: exploring the relationships between the extinction of species and the inventory of life on earth. Zootaxa 1965:61–68
Hoare RJB, Dugdale JS, Edwards ED, Gibbs GW, Patrick BH, Hitchmough RA, Rolfe JR (2017) Conservation status of New Zealand butterflies and moths (Lepidoptera), 2015. New Zealand threat classification series 20. Department of Conservation, Wellington
Hutchings PA, Ponder W (1999) Criteria for assessing and conserving threatened invertebrates. In: Ponder W, Lunney D (eds) The other 99%. The conservation and biodiversity of invertebrates. Royal Zoological Society of New South Wales, Mosman, pp 297–315
IUCN (2012) IUCN red list categories and criteria: Version 3.1, 2nd edn. IUCN, Gland
IUCN (2017) Guidelines for using the IUCN red list categories and criteria. Version 13, IUCN, Gland
Janzen DH (1994) Wildland diversity management in the tropics: where are we now, and where are we going? Vida Silvestre Neotropical 3:3–15
Janzen DH (1997) Wildland biodiversity management in the tropics. In: Reaka-Kudla ML, Wilson DE, Wilson EO (eds) Biodiversity II. Understanding and protecting our biological resources. Joseph Henry Press, Washington, DC, pp 411–431
Janzen DH, Hallwachs W (2016) DNA barcoding the Lepidoptera inventory of a large complex tropical conserved wildland, Area de Conservacion Guanacaste, norhwestern Costa Rica. Genome 59:641–660
Kearney MR, Hossain MA, Sinclair SJ, Song H (2021) Grasshopper country before and after: a resurvey of Ken Key's collecting expeditions in New South Wales, Australia, 70 years on. Austral Entomol 60:52–65
Kudrna O (1986) Butterflies of Europe, Aspects of the conservation of butterflies in Europe, vol 8. AULA-Verlag, Wiesbaden
Leandro C, Jay-Robert P, Vergnes A (2017) Bias and perspectives in insect conservation: a European scale analysis. Biol Conserv 215:213–224
Leather SR (2009) Taxonomic chauvinism threatens the future of entomology. Biologist 56:10–13
LeBuhn G, Droege S, Connor WF, Gemmill-Herren B, Potts SG et al (2013) Detecting insect pollinator declines on regional and global scales. Conserv Biol 27:113–120
LeBuhn G, Droege S, Connor WF, Gemmill-Herren B, Potts SG et al (2015) Evidence-based conservation: reply to Tepedino et al. Conserv Biol 29:283–285
Lewis OT, Senior MJM (2011) Assessing conservation status and trends for the world's butterflies: the sampled red list index approach. J Insect Conserv 15:121–128
May RM (2004) Tomorrow's taxonomy: collecting new species in the field will remain the rate-limiting step. Philos Trans R Soc B 359. https://doi.org/10.1098/rstb.2003.1455
Moir M (2015) Conservation of Australia's plant-dwelling invertebrates in a changing climate. In: Palutikof JP, Boulter SL, Barnett J, Rissik D (eds) Applied studies in climate adaptation. Wiley, Chichester, pp 107–116
Moir ML, Brennan KEC (2020) Incorporating coextinction in threat assessments and policy will rapidly improve the accuracy of threatened species lists. Biol Conserv 249:108715
Moir ML, Hughes L, Vesk PA, Leng MC (2014) Which host-dependent insects are most prone to coextinction under changed climates? Ecol Evol 4:1295–1312
Molloy J, Bell B, Clout M, de Lange P, Gibbs G et al (2002) Classifying species according to threat of extinction. A system for New Zealand. Threatened species occasional publication 22. Department of Conservation, Wellington
Montgomery GA, Belitz MW, Guralnick RP, Tingley MW (2021) Standards and best practices for monitoring and benchmarking insects. Front Ecol Evol 8:579193. https://doi.org/10.3389/fevo.2020.579193
New TR (2007) Broadening benefits to insects from wider conservation agendas. In: Stewart AJA, New TR, Lewis OT (eds) Insect conservation biology. CABI, Wallingford, pp 301–321
New TR (2008) Legislative inconsistencies and species conservation status: understanding or confusion? The case of *Riekoperla darlingtoni* (Plecoptera) in Australia. J Insect Conserv 12: 1–2

New TR, Sands DPA (2002) Narrow-range endemicity and conservation status: interpretations for Australian butterflies. Invertebr Syst 16:665–670

Patrick BH, Dugdale JS (2000) Conservation status of the New Zealand Lepidoptera. Department of Conservation, Wellington

Pollard E, Eversham BC (1995) Butterfly monitoring 2 - interpreting the changes. In: Pullin AS (ed) Ecology and conservation of butterflies. Chapman & Hall, London, pp 23–36

Salvador RB, Tomotani BM, O'Donnell KL, Cavallari DC, Tomotani JV, Saklmaon RA, Kasper J (2021) Invertebrates in science communication: confronting scientists' practices and the public's expectations. Front Environ Sci 9:606416. https://doi.org/10.3389/fonvs.2021.606416

Sands DPA, New TR (2002) The action plan for Australian butterflies. Environment Australia, Canberra

Sands DPA, New TR (2003) Coordinated invertebrate surveys in Australia's National Parks: an important tool in refining invertebrate conservation management. Rec South Aust Mus Monogr Ser 7:203–207

Schuldt A, Assmann T (2010) Invertebrate diversity and national responsibility for species conservation across Europe – a multi-taxon approach. Biol Conserv 143:2747–2756

Scotland RW, Olmstead R, Bennett JR (2003) Phylogeny reconstruction: the role of morphology. Syst Biol 52:539–548

Shaw MR, Hochberg ME (2001) The neglect of parasitic Hymenoptera in insect conservation strategies: the British fauna as a prime example. J Insect Conserv 5:253–263

Taylor RW (1976) Submission to the Australian Senate standing committee on social environment inquiry into the impact on the Australian environment of the current woodchip industry programme. Australian Senate Official Hansard, Canberra, pp 3724–3731

Taylor RW (1983) Descriptive taxonomy: past, present, and future. In: Highley E, Taylor RW (eds) Australian systematic entomology: a bicentenary perspective. CSIRO Publishing, Melbourne, pp 93–134

Tepedino VJ, Durham S, Cameron SA, Goodell K (2015) Documenting bee decline or squandering scarce resources? Conserv Biol 29:280–282

Townsend AJ, de Lange PJ, Duffy CAJ, Miskelly CM, Molloy J, Norton DA (2008) New Zealand threat classification system manual. Department of Conservation, Wellington

Trewick SA (2021) A new species of large *Hemiandrus* ground weta (Orthoptera: Anostostomatidae) from North Island, New Zealand. Zootaxa 4942(2):207–218

Van Swaay C, Warren MS (1999) Red data book of European butterflies (Rhopalocera), Nature and Environment series 99. Council for Europe, Strasbourg

Van Swaay CAM, Warren MS (2006) Prime butterfly areas of Europe: an initial selection of priority areas for conservation. J Insect Conserv 10:5–11

Wilson EO (2003) The encyclopedia of life. Trends Ecol Evol 18:77–80

Chapter 3
Assessing and Understanding Insect Diversity

3.1 Introduction

The reliable measurement of insect diversity, most commonly undertaken as alpha-diversity, depends on consistent recognition of the entities (putatively, 'species') being surveyed. Efforts to contribute valid estimates of diversity of many insect groups have included numerous attempts to avoid or overcome the restrictions of the taxonomic impediment and to obtain information of practical value from less cost and effort than by working from named or nameable species alone. The term 'rapid biodiversity assessment' has proliferated in references to invertebrate surveys in recent decades, and reflects the need to overcome inadequate knowledge and limited logistic support, predominantly by focus on two major themes: some form of taxonomic approximation and the use of selected higher taxa as substitutes (surrogates) for wider taxon representation or changes. Both have been explored extensively. In addition, attention to 'ecological diversity' reflects the numerous and vital roles that insects play in terrestrial and freshwater communities, so that 'taxonomic diversity' is necessarily accompanied by 'functional diversity' in assessing wider conservation significance and the latter become a measure of ecosystem function and sustainability.

3.2 Taxonomic Approximation

Taxonomic approximation, avoiding the routine need for formal species-level diagnoses, itself has two main approaches. First, analyses of samples only to some taxonomic level above species, often only to family and in some studies only to the rather broad level of order and, second, the individual apparently consistent entities within each higher group given some form of numerical or other 'code' to enable recognition. The entities, commonly referred to as 'morphospecies', then

© The Author(s), under exclusive license to Springer Nature Switzerland AG 2022

T. R. New, *Insect Diversity, Declines and Conservation in Australia*, Fascinating Life Sciences, https://doi.org/10.1007/978-3-030-90134-9_3

become the units for assessment, their value depending on consistent allocations but obviating need for formal binomial names. The approach is advocated widely for surveys in which taxonomic expertise is limited, and for taxonomic groups in which taxa can be distinguished on relatively 'obvious' features and so are amenable to survey by people with only limited taxonomic training—these workers have been given various labels, such as 'biodiversity technicians' or 'parataxonomists'. The latter term was developed in conjunction with surveys of tropical insects, as a vital component of enhancing the very restricted specialist taxonomic expertise available to process large and diverse insect samples, mainly from forests. Most parataxonomists were local trained people who collect/sample insects, prepare them, sort the specimens to morphospecies, and assemble the data into spreadsheets or other form. Discussed by Basset et al. (2004), the multiple advantages of working with local parataxonomists in tropical insect faunas included (1) increased sampling efficiency, including year-round field activity (a marked contrast to the short seasonal visits usual for visiting specialists); (2) efficient specimen preparation at low cost; (3) increased integration of local ecological information associated with the specimens; and (4) enhanced local interest and public outreach in 'biodiversity'. However, these steps do not obviate the reality of taxonomic and biological ignorance, even though helping substantially to demonstrate some form of 'species numbers' contributing to an inventory. A major benefit of exercises such as this is to establish a pool of local expertise and interest.

However, some surveys for quality control have revealed serious reservations if relying wholly on this strategy (Cranston and Hillman 1992) and validation of a complete voucher specimen set of all morphospecies by relevant specialists is highly desirable. Confusions or errors from non-experienced workers can flow from differing observer capability as well as factors such as intrinsic variability in size and colour of insects, different metamorphic stages, sexual dimorphism, seasonal variations and others, each of which may lead to over- or under-estimation of richness and inconsistency of specimen allocation. With adequate care, the numbers and changes in relative abundance of morphospecies within a study can provide valid information. In faunas in which many insect taxa are undescribed or difficult to identify reliably, this approach is largely inevitable.

Nevertheless, 'lumping' of species (or morphospecies) uncritically within higher levels (genera, families, orders) inevitably combines units with different biological features or idiosyncrasies—so that it may not be particularly useful to find that one sample (site, occasion) has 'X morphospecies' of beetles or flies while another has 'Y morphospecies'; even if 'Z morphospecies' occur in both samples, no information can be inferred on their significance, trophic roles or wider biology other than in very general terms. Many workers have noted that interpretation of insect samples to order or family level alone has severe limitations beyond a simple enumeration of putative infracategory units present. As one example, a survey of Diptera from habitats across the Atherton Tablelands, Queensland, was based on very restricted sampling at one time of the year, and flies were identified only to family level (Smith and Mayfield 2015): the 172 species found, itself impressive from a short survey, is unlikely to fully represent the array of Diptera present on those sites.

Difficulties arise, also, in attempts to compare surveys by different workers, or across sites and times—unless a well-curated voucher series is available for reference, with this preferably stored centrally in an institution to assure access. Biases from different sampling methods or intensity, and different sorting priorities and approaches, as well as different aims of surveys can all thwart strict comparisons. However, responsible voucher deposition is not always undertaken and, indeed, some potential depositories are reluctant to store such accumulations for lack of space or defined use, or lack of curatorial capability. In practice, many such accumulations are simply discarded at the end of a study or career, rather than archived for future use as methods and expertise develop. Thus, referring to the progressive incorporation of DNA barcode methods in insect inventory studies (p. 76), Miller et al. (2016) commented that 'every bug on a pin or in a deep freeze is a time-stamped genome'. The ideal outcome, of course, is that the voucher material will eventually be studied fully, and morphospecies translated to real (named or clearly diagnosed and consistently recognisable) species.

3.3 Surrogate Taxa

The second theme, involving potential surrogacy, has wide applications in conservation with development of the concepts attending 'biological indicators' (for example, McGeoch 1998, 2007), through which changes in one taxonomic group (or few groups) are taken to reflect parallel changes in others that have not been assessed directly or in equivalent detail. Focused study of such selected key groups may then supplant need for total identifications of all insects found. Claims for such validity are widespread and whilst some are simply 'wishful thinking', detailed investigations continue, with greatest reward when species richness in a studied taxon indeed reflects that in others. Many insect groups and contexts have been proposed to show that some insect (or other arthropod) taxa have bioindicator values. Only from faunas in which the biology of the candidate taxa is reasonably well understood can any functional extrapolation be pursued. Testing the potential value(s) of any suggested bioindicator group is complex, and the outcomes may be very inconsistent. The twin contexts are (1) that richness of a focal taxon reflects richness in nominated other co-occurring taxa and (2) that changes in richness of a focal taxon indicate changes in such other taxa.

Using higher taxonomic levels as surrogates for species-level assessment is an attractive 'shortcut' in insect biodiversity research (Rosser 2017) but such exercises need very careful validation. It is obviously far easier (that is, cheaper and less dependent on specialist involvement) to evaluate genera, families, or even orders than species (or morphospecies), but successively higher levels lose sensitivity—and it is perhaps very little value to know that a sample contains, for example, simply Hymenoptera and Coleoptera. The required level of interpretation ('taxonomic sufficiency') can differ across studies with different purposes, and numerous tests of surrogacy have given inconsistent or complementary outcomes. Some surveys have led to conclusion that family richness can predict species richness, while others have concluded that even genera do not achieve this. Testing several data-sets from surveys of beetles and ants (as among the most frequently studied insects in terrestrial biodiversity surveys), Rosser (2017) found higher taxa unsuitable as surrogates for patterns of species composition, and urged that they be used only with caution. Only when the species:higher taxon ratio is very low may this surrogacy 'work'. In other circumstances stochastic factors render any predictions difficult. HIgher taxon categories can be useful as surrogates in identifying some community changes due to disturbance. For moorland sites with a range of fire histories in Tasmania, this approach was influenced by taxonomic diversity and community structure but both family- and order-level analyses could be effective surrogates for species delimitations (Driessen and Kirkpatrick 2019). In that series of sites in lowland and montane button grass moorland, invertebrates collected by pitfall trapping (so, ground-active) and sweepnetting (foliage-active) were analysed in an unusually broad approach, involving 21 orders, 197 families and 751 'species'. The overall level of site disturbance influenced validity of higher taxon use—in this study, order-level separations performed better in lowland sites on which fire had greater impact on the structure of invertebrate assemblages.

However, Timms et al. (2013) noted that numerous studies in freshwater systems supported the use of higher taxa in interpreting community data, and that more studies on terrestrial arthropods were needed. Their synthesis of nine terrestrial arthropod studies in general supported that coarser taxonomic resolution did not substantially affect how the community composition was interpreted from the species-level data, and that such higher levels can be used validly in RBA. Nevertheless, Timms et al. acknowledged that species-level information is necessary for inventory studies and to address many ecological questions. A significant caveat from their survey was that rather than the inference that species-level information may not be needed for interpretation, that inference may simply reflect that many multivariate analyses are not sufficiently sensitive to detect the subtle information that species data can provide.

As Austin (1999) emphasised earlier, the assumptions underlying any attempt to use surrogate taxa as a shortcut in surveys are valid only if they are tested rigorously—and such research necessarily focuses on the species level, so that accurately distinguishing the taxa is critical. Two kinds of potential errors occur, leading to either inflation or underestimation of the species richness present. Thus, for either context the number of morphospecies designated does not match the number of biological species present: consequently, the presence of cryptic species (p. 104) leads to underdetection or of polymorphic species, determined by the identifier(s), for example through lack of specialist knowledge or experience with the group: the extent of the error is likely to reflect training, knowledge of the literature and previous experience. Although some advocates of such RBA approaches by using less experienced technicians to undertake sorting and categorisation of insect samples have asserted that errors tend to be low and do not confound later analysis (Oliver and Beattie 1993, 1996), others have urged greater caution, and some critical surveys (such as by Wilkie et al. 1999) have demonstrated need for careful monitoring of 'sorter performance' to assure consistency over time.

Whatever approach is used, accurate enumeration of entities in samples of insects from either terrestrial or aquatic environments is often a complex exercise. Long lists of species, or high numbers of morphospecies, are impressive in demonstrating scale of complexity and diversity, but also emphasise the impracticability of holistic assessments and pursuing management. A further complication, often overlooked in practice, is the incidence of cryptic species complexes that are morphologically (almost) indistinguishable but have clearly different molecular identities when their DNA profiles are studied (p. 104). Cryptic species complexes include undiagnosed species with very different biologies masquerading under the same name and with lack of distinction perhaps having serious practical economic ramifications—as in the case of pest (or potential pest) taxa. Superficial attempts to identify species may indeed unwittingly promote dissemination of 'misinformation'.

The need for taxonomic accuracy largely restricts insect species inventories to the best documented taxonomic groups and faunas, with wide acknowledgment that other 'all-taxon biological inventories' (ATBIs) of insects are highly impracticable. Elsewhere, levels of approximation and uncertainty increase. In general, much of the information potentially furnished from ATBI attempts is never released, simply because in many studies the most predominant insect constituents cannot be analysed soundly. Vast numbers of specimens accumulated must be examined critically to detect the rarer and more unusual (low abundance) taxa present. For 'indicator groups', again most evident amongst better documented faunas (such as Europe: Maes and Van Dyck 2005), reliable species-level recognition is a common condition of use, augmented by knowledge of distributions and 'what they are indicating'. The last incorporates knowledge of how they respond to any given environmental change(s), such as tolerance to changed physical, chemical or biological conditions. The species should also be easily detectable and amenable to sampling by standard replicable methods that can be applied without confusion (Samways et al. 2010). In brief, any valid 'shortcuts' or other means to streamllne interpretation of insect surveys for assessing richness or diversity are welcome—and include taxon restriction (as above), and methods of character recording, information storage and retrieval, with increasingly reliable databases that can increase the nexus between taxonomic and ecological information (Didham et al. 2013; La Salle et al. 2009). Many such efforts are driven by needs to overcome the taxonomic impediment and (as La Salle et al., 2009, put it) for 'fundamental changes to the taxonomic process and paradigm to increase taxonomic productivity by orders of magnitude'. However, 'We cannot lose sight of the fact that the purpose of a field inventory is both to know what is there and what it does, and to generate comparative information beyond the inventory' (Miller et al. 2016).

However rewarding and tangible when it succeeds, it is impossible to base practical insect conservation solely on the measures taken for individual species (New 2009). Following the example from Odonata (Clausnitzer et al. 2009), sensible wider conservation measures projected to collectively conserve much of the fauna in selected areas—so, for example, including the most threatened forest species of lowland tropical habitats—may be feasible (Table 3.1) and move progressively toward increased ecological representation across major ecosystems or biotopes.

Table 3.1 Some general conservation measures proposed to cover suites of threatened species rather than preparing action plans for each species individually: measures suggested for Odonata by Clausnitzer et al. (2009)

1. Prevention of any further deforestation.
2. Afforestation with indigenous trees
3. Fire control
4. Leaving a strip of natural riparian vegetation of at least 20 m on either side of streams and rivers in cultivated or mined landscapes
5. Water pollution control measures
6. Protection of watersheds, floodplains, water retention sites, swamps, etc., against adverse impacts of damming and other alterations

3.4 Ecological Diversity

Although some estimate of taxonomic diversity is a predominant concern in much insect conservation activity, and is the foundation for appraising 'diversity' and detecting declines or simply defining the current state as a 'single 'snapshot' or as part of a series of observations transcending sites or across times, this is fundamentally a picture of community or assemblage structure alone, without further implications for ecological functions. Allied with this, the ecological or functional diversity of the insects present directly affects the integrity of the communities, and its maintenance is a major concern in conserving ecological integrity and sustaining the 'ecological services' provided. Thus, representative trends such as (1) declines of pollinating insects; (2) changes in insect-plant associations that may reflect floristic changes; (3) changes to food web structures from altered imbalances between trophic guilds; and (4) particular traits (such as body size, dispersal capability, voltinism, and others) may affect vulnerability to particular environmental changes may induce a more generally-perceived outcome that 'ecological specialists' may become more susceptible than more generalist species with wider ecological tolerances. The loss of insect diversity inevitably entails changes to interactions within the community and possible alterations to its stability and persistence.

As representative extremes, (1) a truly monophagous insect cannot persist without its specific host plant or prey and (2) a plant with a single insect pollinator species cannot continue without that pollinator; mutual dependencies between species are diverse, and their dynamics not fully understood (New 2017). Many interactions involving insects are more flexible, but demonstrate the need to include ecological or functional diversity in conservation assessments, with realisation that disruptions of particular interactions can lead to 'cascade effects' extending through food webs and community structure. Insect species span the full range from influential 'keystone species' whose loss may be calamitous to many that are presumed unimportant or 'redundant' in the context of ecosystem function. The precise roles and resource specificity of most insects present remain unknown, other than allocation to some broad trophic group. It is thus also unclear whether their loss will have wider deleterious effects, or whether that loss can be compensated by other species present so that ecosystem functions are unaffected. Defining species that are indeed 'functionally redundant' is uncertain, not least because rare species can be functionally important and the roles of individual species change with changing community composition and with their abundance. Some insects that normally occur in low numbers undergo sporadic or periodic outbreaks: some stick insects in southern Australian forests occasionally appear in vast numbers, leading to much increased defoliation, and some grasshoppers may undergo similar increases to devastate crops of grasslands. For most of the time, those insects are unnoticed.

However, if some species are indeed functionally redundant, from some viewpoints they become expendable, notwithstanding any evolutionary or taxonomic interest or their conservation status. Even if not redundant, rare species of conservation significance may simply be too scarce to affect community composition or integrity, and arguing for their conservation on grounds of ecological importance becomes misleading. Lyons et al. (2005) noted that the ecological effects of rare species are largely unknown, but that describing their functional roles becomes important in clarifying whether diversity and ecosystem fitness are indeed correlated.

Despite widespread traditional belief that high levels of biodiversity (as species richness) are correlated with high levels of ecosystem function, this is difficult to ascertain—but is powerful argument for the conservation of biodiversity. If, conversely, ecosystem functions can be maintained with lower species richness, arguing for ecosystem functions does not also oblige conservation of all species present. Precautionary conservation measures are needed to counter the uncertainties, largely reflecting that the roles of most insect species are unknown. As Schwartz et al. (2000) put it 'the observation of a positive link between diversity and function . . . is neither simple nor universal', with very different inferences emerging from various studies. Two contrasting scenarios (Fig. 3.1) allow for many intermediate situations. Adequate precautionary measures are needed on the assumption that all species contribute in some way to the well-being of the ecosystems in which they occur.

One extreme scenario that could occur is of ecologically dominant alien species fulfilling entirely the roles of native species. In such a situation, with all ecological functions satisfied by alien species alone, all native species become redundant and expendable—a sentiment abhorrent to conservation interests focused on other parameters, but perhaps already approached by a few generalist pollinator bees of concern over implications of displacing native pollinators. They could be lost without impairing ecosystem functions (Schwartz et al. 2000). Slightly more broadly but equally concerning for conservation of native insects and other taxa, Schwarz et al. warned that—hypothetically—in terms of sustaining functions, outcomes such as 'we could make a list of ten integral species for a given ecosystem and dispense with the remaining diversity' might arise. The large numbers of insect species in many terrestrial communities, in particular, render it difficult to determine which species have any functional significance, and the detailed manipulative experiments needed to explore this are also very difficult to pursue. The approach of assessing 'structural redundancy' in species compositions may be helpful in assessing functional redundancy (Truxa and Fiedler 2016). Their surveys of the larval feeding

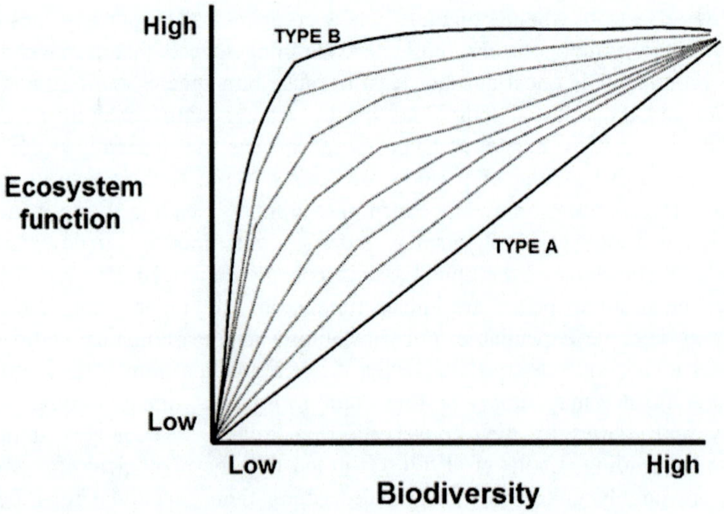

Fig. 3.1 Hypothetical relationships between biodiversity and ecosystem function, where a positive correlation between these exists. Type A, as linear, expresses a condition in which all species contribute to function. In contrast, Type B illustrates the situation in which only few species contribute to ecosystem function, so that the others are redundant or too scarce to influence those functions. Intermediate lines exemplify the numerous intermediate cases between these extremes (after Schwartz et al. 2000)

habits of floodplain moths in Austria suggested how the 12 functional types found might also indicate strong functional redundancies.

References

Austin AD (1999) The importance of 'species' in biodiversity studies: lessons from a mega-diverse group - the parasitic Hymenoptera. In: Ponder W, Lunney D (eds) The other 99%. The conservation and biodiversity of invertebrates. Royal Zoological Society of New South Wales, Mosman, pp 159–165

Basset Y, Novotny V, Miller SE, Weiblen GD, Missa O, Stewart AJA (2004) Conservation and biological monitoring of tropical forests: the role of parataxonomists. J Appl Ecol 41:163–174

Clausnitzer V, Kalkman VJ, Ram M, Collen B, Baillie JEM et al (2009) Odonata enter the biodiversity crisis debate: the first global assessment of an insect group. Biol Conserv. https://doi.org/10.1016/j.biocon.2009.03.028

Cranston PS, Hillman T (1992) Rapid assessment of biodiversity usng 'biological diversity technicians'. Aust Biol 5:144–154

Didham RK, Edwards OR, Leather SR, Basset Y (2013) Arthropod diversity and the future of all-taxa inventories. Insect Conserv Divers 6:1–4

Driessen MM, Kirkpatrick JB (2019) Higher taxa can be effective surrogates for species-level data in detecting changes in invertebrate assemblage structure due to disturbance: a case study using a broad range of orders. Austral Entomol 58:361–369

La Salle J, Wheeler Q, Jackway P, Winterton S, Hobern D, Lovel D (2009) Accelerating taxonomic discovery through automated character extraction. Zootaxa 2117:43–55
Lyons KG, Brigham CA, Traut BH, Schwartz M (2005) Rare species and ecosystem functioning. Conserv Biol 19:1019–1024
Maes D, Van Dyck H (2005) Habitat quality and biodiversity indicator performance of a threatened butterfly versus a multispecies group for wet heathlands in Belgium. Biol Conserv 123:177–187
McGeoch MA (1998) The selection, testing and application of terrestrial insects as bioindicators. Biol Rev 73:181–201
McGeoch MA (2007) Insects and bioindication: theory and progress. In: Stewart AJA, New TR, Lewis OT (eds) Insect conservation biology. CABI, Wallingford, pp 144–174
Miller SE, Hausmann A, Hallwachs W, Janzen DH (2016) Advancing taxonomy and bioinventories with DNA barcodes. Philos Trans R Soc B 371:20150339
New TR (2009) Insect species conservation. Cambridge University Press, Cambridge
New TR (2017) Mutualisms and insect conservation. Springer, Cham
Oliver I, Beattie AJ (1993) A possible method for the rapid assessment of biodiversity. Conserv Biol 7:562–568
Oliver I, Beattie AJ (1996) Invertebrate morphospecies as surrogates for species: a case study. Conserv Biol 10:99–109
Rosser N (2017) Shortcuts in biodiversity research: what determines the performance of higher taxa as surrogates for species? Ecol Evol 7:2595–2603
Samways MA, McGeoch MA, New TR (2010) Insect conservation. A handbook of approaches and methods. Oxford University Press, Oxford
Schwartz MW, Brigham CA, Hoeksema JD, Lyons KG, Mills MH, van Mantgem PJ (2000) Linking biodiversity to ecosystem function: implications for conservation ecology. Oecologia 122:297–305
Smith TJ, Mayfield MM (2015) Diptera species and functional diversity across tropical Australian countryside landscapes. Biol Conserv 191:436–443
Timms L, Bowden JJ, Summerville KS, Buddle CM (2013) Does species-level resolution matter? Taxonomic sufficiency in terrestrial arthropod biodiversity studies. Insect Conserv Divers 6: 453–462
Truxa C, Fiedler K (2016) Massive structural redundancies in species composition patterns of floodplain forest moths. Ecography 39:253–260
Wilkie L, Cassis G, Gray M (1999) Quality control on invertebrate biodiversity data compilations. In: Ponder W, Lunney D (eds) The other 99%. The conservation and biodiversity of invertebrates. Royal Zoological Society of New South Wales, Mosman, pp 147–153

Chapter 4
Improving Understanding of Insect Diversity

4.1 Introduction

The various dilemmas and inadequacies of assessing insect declines and conservation needs are not insoluble. However, they must be recognised and acknowledged, together with realisation that conservation must proceed without the full background of understanding needed. A series of eight 'no regret' measures (to be undertaken without full understanding of their likely impacts on insects) were recommended for immediate action by Harvey et al. (2020), and are mirrored in the general measures suggested for Australia and elsewhere (p. 165). They reflect the diversity of approaches and stakeholder participants that can contribute to practical conservation (Table 4.1). The overall aim of these measures is to 'slow or stop insect declines'. Some address single threats or threat complexes, whilst others are more embracing. All, however, focus on themes that can be understood realistically by participants, and all offer potential tangible benefits for human welfare, as well as for the insects themselves, and extend the scope of insect conservation. Harvey et al. (2020) also recommended large-scale assessments of the conservation status of insect groups as short-term needs in order to define priority species, areas and issues, and to increase the numbers of IUCN Red List assessments, all adding to the basic information on which management can be founded. These urgent actions were to be accompanied by (1) medium-term new research to distinguish the effects of different factors driving insect declines, and continuing to analyse data on insect diversity (including that for areas that are currently underexplored) from all available sources, and (2) long-term actions of forming partnerships with the aims of protecting, restoring and creating insect habitats, managing key threats, and promoting and extending standardised monitoring protocols.

At least in theory, many discussions of putative threats and their impacts on insects have included suggestions for threat decrease or remediation of consequences through practical management. Not all those suggestions are immediately

© The Author(s), under exclusive license to Springer Nature Switzerland AG 2022

T. R. New, *Insect Diversity, Declines and Conservation in Australia*, Fascinating Life Sciences, https://doi.org/10.1007/978-3-030-90134-9_4

Table 4.1 The eight 'no-regret solutions urged for immediate implementation to slow or halt insect declines (Harvey et al. 2020)

1. Increase landscape heterogeneity in agriculture
2. Reduce light, water and noise pollution
3. Phase out pesticide use, and replace with ecological measures
4. Reduce imports of ecologically harmful products
5. Avoid and mitigate alien species introductions
6. Conservation of threatened species
7. Enhance restoration and conservation programmes
8. Education for awareness, citizen science and capacity building

practicable, and many require considerable increase in resources to prosecute them effectively. Increased effectiveness depends on increased information, gaining of which in turn may depend on public and political support. Without that support, the shortcomings of the public and political dilemmas persist; appreciation of the values of insects and the needs to conserve insect diversity is a core requirement for all stakeholders, and may involve 'marketing' insects to people who know little about them and, indeed, may fear them or be predisposed to 'swat' them. Insects are 'newsworthy', and publicising novel findings, or peculiarities of their biology, has been a widespread benefit in eliciting interest in species conservation programmes.

Progress is increasing toward including insects in wider conservation agendas as active components rather than simple passengers (New 2007), not least because their importance in wider ecological contexts can be displayed. For both the general public and influential policymakers, evidence of insect importance in sustaining ecosystems is a pragmatic component of communicating their values to humanity. Including known threatened insects routinely in environmental impact statements needed to allow developments can become an avenue to increasing functional biological knowledge from developer investment rather than from public expenditure.

Seeking maximum benefit from insect species conservation programmes, and to insects from management for other biota, rather than simply seeking 'passive passenger benefits' can be augmented considerably by prior planning and consultation. Each focal species has potential to provide some form of umbrella benefits to others sharing its biotope or resources, and awareness of any other 'listed' taxa co-occurring with it can increase priority and importance for the sites involved. An allied theme, championed for threatened Lepidoptera in Britain through 'Butterfly Conservation' (Ellis et al. 2012), is to recognise the importance of metapopulations through promoting landscape-level conservation planning, whereby networks of habitat patches are targetted across a wider landscape. The approach was predicated on recognition that species' declines in Britain occurred for four main reasons, as (1) many protected areas were too small to prevent random fluctuations from driving small populations to extinction; (2) many habitats were represented only by tiny remnants; (3) many surviving patches of semi-natural habitats were either unmanaged or poorly/inadequately managed; and (4) many surviving sites were

Table 4.2 Lessons for landscape-scale conservation of insects from 12 key case studies of butterflies in Britain (from Ellis et al. 2012)

1. Species conservation can be very effective at a landscape scale
2. Careful targetting is essential to maximise success of a project across a landscape
3. Extinction of species on small isolated sites within a landscape need not be inevitable if they are properly managed
4. The principles of landscape-scale conservation can be applied successfully at a relatively small scale and are relevant even at the site level
5. Larger-scale projects provide the opportunity to support and restore a traditional low intensity or an innovative management system that might otherwise not be commercially viable
6. Skilled project officers are essential
7. Land managers have been very receptive to landscape-scale projects
8. Experienced and well supported contractors are essential
9. Local volunteers can play a significant role in landscape-level conservation
10. Development and implementation of landscape-scale projects must be underpinned by sound ecological research
11. Good quality spatial data is essential for project design and delivery
12. The establishment of suitable monitoring systems is essential to assess the effectiveness of the project
13. Habitat condition assessments should also be undertaken to analyse changes to habitat quality
14. Butterflies and moths respond rapidly to changes in habitat quality
15. Landscape-scale projects focused on a single butterfly can benefit a suite of other species which have broadly similar habitat requirements
16. All our landscape-scale projects could only be delivered with adequate medium to long-term funding
17. Well-designed agri-environment and woodland grant schemes are a key delivery mechanism for landscape-scale conservation
18. The maintenance of existing high-quality habitat is more cost effective in the long run than restoration management
19. Landscape-scale conservation always involves partnership working
20. Publicising landscape-scale projects is important for both partner organisations and funders, but also for local communities

functionally isolated within otherwise inhospitable landscapes. Following recommendations in a wider report (Lawton et al. 2010), the perceived need was for 'more, bigger, better managed and joined up' sites. Pursuing this target through species-focussed management endorsed the importance of species-level conservation to which people can relate easily, and a review of 12 such studies in Britain highlights a number of lessons that have much wider relevance (Table 4.2), and endorse the benefits of this wider perspective. The five broad categories of lessons are complementary, and collectively demonstrate the range of relevant contexts and outcomes—but with a common plank of effective communication and supervision by well-informed project officers in the context of partnerships with stakeholders and community groups.

All of the seven 'impediments' discussed earlier need serious efforts for them to be to overcome. Redressing the scientific dilemma (p. 40) links strongly with the Linnean shortfall, namely the need for more taxonomists and taxonomic work, including succession planning to assure continuity of interests and expertise in diagnosing and describing species. It is easy, and a common misconception, for policymakers not to appreciate the practicalities arising from vast numbers of species and the complexity of interpreting diverse samples of insects from ecological surveys and identifying species of conservation interest. In some cases, their reactions devolve on sentiments such as 'if it hasn't got a name it can't be important, so why waste time and effort on conserving it?'. These reactions are difficult to overcome, but can provide opportunity to initiate discussions of the real extent and complexity of insect diversity. Simply clarifying the various meanings of 'rarity' (as a fundamental but often emotive term) in conservation, and its parallels with the three shortfalls of Cardoso et al. (2011)—namely that rarity has dimensions of small distribution (linking with the Wallacean shortfall), low abundance (considered in the Prestonian shortfall) and ecological ambit (the Hutchinsonian shortfall) and as displayed by Rabinowitz (1981) (Fig. 4.1)—helps to understand why threats occur, and their relative impacts on different insect taxa. Rarity does not necessarily enhance extinction risk as is often assumed, but those species with any or all of having small distributions, occurring in low abundance and needing specialised ecological conditions may indeed be more vulnerable than others if environmental changes occur, so that 'This makes rarity a good proxy for extinction risk' (Fattorini

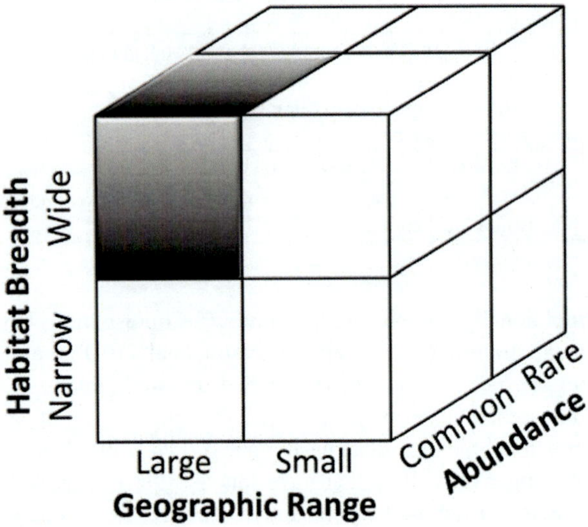

Fig. 4.1 The three dimensions of rarity recognised by Rabinowitz (1981) that combine to form seven categories of rarity, based on habitat specialisation, geographical range and abundance. These three axes are shown as gradients of a rectangle, each divided into two components; one eighth of the divided block (shaded) represents 'common', and the other seven represent some form of 'rare' because of different conditions of the criteria; the most extreme form of rarity is a combination of habitat specialisation, small geographical range and low abundance. All three may predispose a population or species to vulnerability

et al. 2013), whilst the concepts of rarity, vulnerability and extinction risk may need to be distinguished carefully.

4.2 Understanding Species

A universal need is to try to reduce ambiguities from enumerating species, by addressing the complexity of the apparently simple question of 'what is an insect species?' Although many insect species are indeed very distinct, many others are intrinsically difficult to diagnose and distinguish unambiguously, often due to very subtle structural or biological peculiarities, or to the different species concepts and methods for analysis that have been advanced. Nevertheless, a broader suite of semantic problems have been discussed, influencing the use of 'species' as the major—presumed tangible and definable—unifying currency in conservation, in reflecting needs to enumerate and recognise components of diversity in ways that can be quantified and compared. Those needs are vital (as numbers or lists of taxa) to managers and others, with implication that those units are recognisable consistently and enable valid comparisons across sites or times. This approach, often accepted uncritically, is indeed attractive, but care may be needed in defining species. A thoughtful discussion by Raposo et al. (2020) points out the difficulties in perceptions that they regard as 'a profound "conceptual turbulence" in the knowledge flux between taxonomy and conservation biology', in that 'in general, taxonomists only vaguely understand what conservationists wish to preserve and conservationists appear not to consider more profound taxonomic issues and the consequences for their interests'. Their discussion focused largely on mammals and birds, with examples of different taxonomic methodologies and philosophies markedly influencing the stated 'numbers of species' and that may render delimitations more difficult for conservation practitioners. For insects, species delimited by DNA barcode analyses (p. 76) render practicalities increasingly difficult—but many 'conventional' insect species (those separated on structural features and amenable to inclusion in illustrated dichotomous keys from which non-specialists may identify the species) already provide far more complexity and ambiguity than can be accommodated easily by many non-specialist users. There is little prospect for the above perception problems being solved, and the major role of 'species' in conservation is to provide footholds for awareness and management. However, it is perhaps also pertinent to remember Charles Darwin's sentiment, reiterated by Raposo et al., that essentially a species is what an experienced naturalist says it is! Consensus in common sense in practice must outweigh the artefacts and ambiguities that can arise from contrasting methodologies and philosophies. However, differences in taxonomic methodology and the differing opinions of specialists ensure that any insect 'species list' or 'species number' is dynamic and a 'work in progress' as an interim expression of belief in the delimitation of the constituent entities. Whilst 'species' will almost certainly continue to be the functional units on which much insect conservation devolves, and from which losses, extinctions and increases are projected, it is important to acknowledge shortcomings of regarding them as static entities. As Raposo et al. (2020) reiterated 'taxa are scientific hypotheses' that are subject to revision and change as further information accrues.

Among insects, presumed intraspecific variations can become very difficult to interpret. Variation amongst individuals of a presumed single species may be normal, or may even reflect occurrence of a suite of very similar (sibling or cryptic) species that differ in genetic or biological features but are indistinguishable by the usual morphological characters that designate species in that taxon. Conversely, 'normal' intraspecific variation may be entirely trivial or be related to seasonal morphs, sexual dimorphism or other polymorphisms (such as in mimicry complexes), or geographical variations. Falsely distinguishing any form of these variations as true species inflates species richness, but even when variations are well-documented their status—such as the effects of different host plant species among herbivorous insects to produce local ecological 'races'—may be debated by specialists who hold very different views but whose opinions may appear equally valid. An unusual situation can occur through hybridisation of forms within a variable species to produce local populations with very distinctive features, which are recognised easily and named to facilitate reference. In eastern Australia the polymorphic Swordgrass brown butterfly (*Tisiphone abeona*, Nymphalidae) has several named local forms and the considerable geographical variation has led to erection of 6–8 'subspecies', based on wing features. The complex has been discussed extensively (Braby et al. 2012). Where the predominant southerly and northerly groups meet in central coastal New South Wales, the variable form 'joanna' represents a consequence of secondary contact between populations, leading to hybridisation in combination with local adaptation, introgression, presumed genetic drift through geographical isolation and limited gene flow.

The long hobbyist interest in, and attention to, butterflies is particularly confusing, and important in debating taxon numbers and the validity of local 'forms' as conservation targets. Many local 'forms' (as above) and unusual varieties of butterflies have been distinguished primarily on their wing colour and pattern, and have been given names to reflect that distinctiveness. In many cases, those names are simply descriptors and not intended to convey taxonomic meaning—but as trinomial epithets are formally 'subspecies' and subject to the same formal rules of close investigation and assessment to confirm their true status. Some ambiguities among Australian butterflies were discussed by New (1999) and the wider relevance of subspecies in conservation appraised by Braby et al. (2012). Most Australian butterfly subspecies have been erected on their general adult appearance with little, if any, quantitative evaluation of their peculiarities. Many of the butterflies nominated for conservation concern are currently localised 'subspecies', with this level treated as fully equivalent to 'species' in regulations, even though their taxonomic status may not have been defined critically. For full credibility, it becomes important to clarify whether each is indeed an evolutionarily significant unit (ESU). Especially when subspecies are formally 'listed', their credibility becomes paramount in determining whether they receive conservation attention.

As Dennis (1977) noted for British butterflies, a trinomial name properly has one of two meanings—as a formal taxonomic entity and a distinct evolutionary unit within a species, or as a simple descriptor without any intent beyond being a convenient label. However, marked inconsistencies persist in how such names are applied. Braby et al. (2012) provided working criteria for this in butterflies, in terms of subspecies being allopatric, evolving populations of a species that are at least partially isolated, phenotypically distinct and with at least one diagnostic character state, with characters associated with 'evolutionary independence'. In butterflies, relevant phenotype character states include wing colour pattern, morphology, and features of behaviour or ecology, so that allopatric subspecies when defined in this way can help to clarify boundaries between definable taxa. Without such definition, many infraspecific forms are defined only rather subjectively and inconsistently. Braby et al. discussed the approach initiated by Tobias et al. (2010) for distinguishing such taxa in birds, with allocated scores summing to a quantitative index, and for which 'threshold values' for the various characters were difficult to define fully. Ideally, and as Dennis (1977) advocated, a subspecies should be recognisable by some 'saturation differences', features unique to all individuals allocated to it. That distinctiveness may be clear in populations at the extremes of a geographical range, but in some cases the various 'forms' intergrade subtly and continuously as a cline that reflects gradients in environmental features. Assessing the biological reality of subspecies should ideally encompass knowledge of a series of features, without which suggestions of some subjectivity are inevitable. New (1999) listed those features as (1) the patterns of variation within the species concerned;(2) the detailed distribution of the species; (3) the degree of isolation of the various constituent populations, reflecting (4) the dispersal capability of the adults and (5) the biological peculiarities of the populations. Over much of the world, most such information is not likely to be available for insects other than for some butterflies. And, even for most well-studied butterflies, gaps are likely in either or both of phenotypic and genetic variation, and the extents of population isolation unclear, together with whether some 'subspecies' are really 'full species'. Most insects of conservation concern are nevertheless deemed to have largely closed populations on isolated sites within altered landscapes, and as such are sufficiently distinct and potentially separate evolutionary units to merit conservation attention.

Although perhaps most common to think of isolation in terms of islands surrounded by sea, the numerous functional terrestrial analogues include discrete patches of food plants or natural vegetation in agricultural or urban landscapes, mountains or hills surrounded by lowland, patches of forest in cleared areas, swamps or ponds lacking direct continuity with other waterbodies, caves, and many others in which restricted environmental conditions differentiate them from the wider landscapes in which they occur. Some insects have been isolated for a very long time. Many cave insects, for example, have been restricted to isolated caves for millenia without external disturbances and without need or ability to move elsewhere, continuing to thrive and evolve in stable hospitable environments. This situation contrasts markedly with processes such as the rapid clearing of natural vegetation for human needs, in which large proportions of suitable habitat can 'instantly' disappear to leave resident insects only in newly isolated remnant patches. In most instances those taxa have not yet had time to differentiate obviously, and may not survive to do so.

Progressive development of allopatric species or subspecies by evolutionary divergence of a single lineage is a source of confusion in many analyses of insect complexes, reflecting the 'grey zone' of Braby et al. (2012). Thus, some interpretations of cryptic species complexes by DNA barcode analysis can reveal unexpectedly high numbers of entities which may be considered to be 'species'. One classic example is of a Neotropical skipper butterfly (then known as *Astraptes fulgerator*, Hesperiidae), for which Hebert et al. (2004) revealed 10 species masquerading under that name and apparently otherwise differing only in some details of DNA sequences. Those conclusions were queried by others, based on the methodology used, and Brower (2006) made a more robust analysis to show that the likely number of species was 'at least three … but probably no more than six or seven', and a variety of other conclusions have been advanced. Nevertheless, Brower (2010) elected to formally name the 10 originally separated species, simply to provide taxonomic identity and replace 'operational taxonomic units' (OTUs) with names that were more relevant, rather than retaining informal epithets that were essentially *nomina nuda*. The discussion accompanying Brower's action was in part provocative: he wrote 'If publication of the names and associated diagnoses provokes an outcry, so much the better for science' and 'simply because species can be identified and publicised on the basis of a few nucleotide differences, it does not mean that they should be'. However, later the DNA differences were associated with caterpillar colours, foodplant relationships and subtle adult butterfly colour pattern differences to confirm that the 10 taxa (with another later added from more recent discovery) were indeed consistently recognisable and distinct (Janzen and Hallwachs 2016). However, the implications of this approach extend far more widely. A presumed generalist braconid wasp (then known as *Apanteles leucostigmus*) attacking hesperiid caterpillars at Guanacaste (Costa Rica) was revealed by DNA barcoding to comprise 39 species of highly host-specific microgastrine wasps, subsequently distinguished also by small morphological features that had been overlooked or unheeded previously, and each attacking only an individual host species. In parallel, the 16 species of parasitoid tachinid flies were found to comprise 72 species, most of them host-specific and only nine likely generalists. In short, DNA barcoding transformed awareness of insect diversity appraised by more traditional morphology-based taxonomy and—in Guanacaste—revealed many hundreds of species that were previously overlooked and undescribed and helped to elucidate many interactions between those species. Morphological differences among many groups of Diptera and parasitoid Hymenoptera are very subtle, and the recognition of species within complexes has led to the reality that specialist taxonomists may work only on individual genera or, at most, families of these insects. The microgastrine wasps noted above are only one such example. Their complexity at Guanacaste was illustrated by DNA barcoding of 2597 reared wasps to reveal 142 previously unrecognised provisional species, increasing the then known 171 morphologically recognised species to a total of 313 provisional species, most of which are extremely host-specific and attack only one or two host species. The six microgastrine genera evaluated, and for which morphology-based species estimates were available, collectively increased the total species numbers by more than 70% (Smith et al. 2008).

The Guanacaste project has focused on attempting to prepare an inventory of the Lepidoptera present in this area of Costa Rica through associating caterpillars with adults, expanding to identify their foodplants and parasitoids and hence sort out the participants and parameters of an enormous food web based on this diverse group of herbivorous insects. The survey commenced in 1978, and continues, now incorporating tens of thousands of species, including an estimated 15,000 species of Lepidoptera. A summary of the historical and practical background (Janzen and Hallwachs 2016) includes the comment that the most salient outcome of incorporating DNA barcoding into the inventory was 'to realise that it would be a seriously debilitating oversight to conduct a species-level arthropod biodiversity inventory of a large complex tropical biological system' without that approach and its later developments.

More traditionally, the approach has had its critics and doubters. The understandable impetus to seek rapid ways to acclerate species diagnoses and overcome the dearth of specialists can have many such pitfalls. DNA barcode analysis has been described as 'a quick and easy standardised method' without presenting intellectual problems (Stoeckle and Hebert 2008), and the vast scale of possible investigations was noted by Brower (2010) as 'generated in large part by nameless technicians in assembly-line settings'. Interpretation of barcode data to provide satisfactory taxonomic evaluation may depend on the insights from experienced specialists on the group concerned and who can assess the significance of the variety exposed. As Borkent et al. (2018) asserted, morphological studies are far more informative in understanding diversity than species known only by barcodes or 'numbers', and the latter provide no clues to their ecology or biology. Whilst such codings can indicate the extent of diversity and suggest numbers of distinct entities, and provide benefits such as associating the two sexes in samples and adults with larval stages, the ecology of those entities may remain obscure without more direct study. However, Janzen's magisterial and penetrating long-term study in the Area de Conservacion Guanacaste of Costa Rica incorporates (and demonstrates) aspects of insect diversity and ecological relationships that are revolutionising understanding of insect community richness. Not yet paralleled fully for Australia, it is clear that the revelations of DNA barcoding are likely to (1) considerably increase the numbers of insect species recognised from any given site, perhaps especially in the tropics, (2) transform appreciation of diversity to increased awareness of the richness of specialists rather than generalists, and (3) continue to query the 'real identity' of many of the described species distinguished by traditional morphological taxonomy alone. In short, uncertainties over the constitution of insect inventories could become overwhelming. Difficulties of taxon enumeration were signalled by examination of the Psylloidea (Hemiptera) retrieved from only eight nights of light-trapping in Queensland, from which 67 of 72 entities detected by DNA barcoding were probably undescribed species (Martoni et al. 2020), giving a very high ratio of 'undescribed: described' taxa. Applicable also to many other insect groups in Australia, Martoni et al. (1) commented that the fauna is strongly underdocumented and (2) hypothesised that the true number of psylloid species may indeed be vastly higher than those already known.

More widely, harnessing the information available from the world's major institutional insect collections has immense potential to advance taxonomic and distributional knowledge. Jin et al. (2020) commented that those collections 'have long been the principal resource for studies of biodiversity and evolution' as collectively containing 'essentially all species known to science'. Historical specimens in museum collections include many species that are now extinct, rare or generally difficult to obtain, and many are suitable for genomics studies founded in the CO1 DNA barcode region. That information, not always clear because of degeneration, contributes both to (1) revealing additional diversity, sometimes unexpectedly high, in species richness (see above) and (2) helping to clarify phylogenetic relationships that can further define and characterise faunal peculiarities. Two recently investigated Australian examples illustrate the kinds of outcome from this approach that are likely to become more frequent:

1. Cryptic diversity amongst morphologically similar insects is by no means confined to the small-bodied taxa that are conventionally among the least investigated. The longhorn beetle subfamily Prioninae (Cerambycidae) contains the largest wood beetles and some of the largest of all insects—with some species reaching up to 175 mm in length. Adults are generally short-lived, and many apparently do not feed. In contrast, their wood-feeding larvae can take up to about 10 years to develop, and are important in recycling of dead timber. Adults of many species are quite similar in appearance, and until recently the Australian representation of the subfamily comprised 52 species in 24 genera. Examination of DNA profiles using a broad 'Whole Genome Shotgun' sequencing of historical specimens in Australian museums led to recognition of 48 additional species (Jin et al. 2020), so almost doubling the known Australian prioninine fauna, description of six new genera, and some other phylogenetic rearrangements within the subfamily.

   That study also brought out another theme of much wider application in identifying insect species, that some previous synonymies of species names applied to morphologically similar taxa may actually be misleading in falsely amalgamating truly different species.
2. The Ichneumonoidea, an enormous group of parasitoid wasps, has conventionally been divided into two large families, Braconidae and Ichneumonidae. Both are diverse: globally, Quicke et al. (2020) quoted more than 47,000 species and 'undoubtedly many times this number'. The large (around 25 mm body length) wasps comprising a group long known as the Trachypetinae comprise three endemic Australian genera, each with few species, and apparently confined to the interior arid areas. Nothing is known of their biology or host relationships, but at least some species are nocturnal. Their taxonomic position within the Braconidae has for long been anomalous (Austin et al. 1993) but recent DNA analysis led to designation of Trachypetidae as a distinct family as a sister group

to the Braconidae (Quicke et al. 2020), so adding a further endemic family to Australia's insect fauna.

Such examples help to emphasise the importance of insect surveys and proper archiving and storage of the material obtained. Especially in remote areas—well illustrated by much of Australia far from major human settlements—costs of protracted or repeated surveys can become high, often prohibitive, and even single field trips can easily escalate to the status of major expeditions. Funding for travel, equipment, personnel, processing and storage of specimens and so on can rapidly become prohibitive. Gaining maximum benefit from insect surveys, in conjunction with other sources of relevant information, and prior consultation with people likely to advise or participate in interpretations and identifications is strongly advisable, not least in honing the sampling or collection methods from the multitude of options available (Samways et al. 2010). Specialist support and participation in such exercises is a scarce commodity, and should never be 'taken for granted', even if it is available and fundable. Museum curators and senior staff are commonly over-extended.

One important potential source of information on insect incidences and diversity in many parts of the country is the 'bycatch', the mass of non-target insects accumulated during programmes of regular surveillance for pests, or from surveys of individual groups or taxa, and which is commonly discarded without examination, as irrelevant. Not all are immediately obvious—invertebrates from pitfall traps used to survey small mammals or reptiles, for example, can be many and varied. If documented and stored properly, that material can become components of 'ecological collections' (p. 170) for later use, but that archiving may be difficult for understaffed institutions to accommodate: in many cases that material is treated as a liability rather than a 'treasure trove'. Indeed, reflecting limitations on space for indefinite storage and lack of immediate interest for identifications, it is commonly discarded. However, without such materials, the potential for 'snapshot' or interval comparisons of insect assemblages over time may not be possible other than from any material of particular groups retained.

Examinations of bycatch can add substantially to basic documentation of local insect faunas, augmenting inventories and providing new distribution records and even revealing incidences of known threatened species (in which case it may become necessary to review the use of any continuing surveillance and insect trapping, as constituting a possible threat) and—as for other survey methods—previously unknown taxa. Nevertheless, as Hribar (2020) noted, bycatch can contribute substantially to basic entomological documentation for inventory, species incidences and distribution mapping. His example, from examinations of 21 years of bycatch from mosquito surveillance in Florida, generated many valuable new records, and at least eight new species, and were a notable contribution to increasing knowledge of the state's insect fauna. Logistic problems of examining such material, from Hribar's example, can include the sheer numbers involved (with mosquito numbers 'in the tens of thousands per trap night') as well as sorting and preparing the residue and, importantly, the interest, time, and funding to do so. Non-target catches, not always retained or in a condition fit for detailed examination, are very common in pest management measures and, as in survey bycatch, numbers killed can become enormous. An early study of use of electric insect traps ('zappers') by householders to reduce incidence of mosquito bites in the United States yielded 104 non-target insect families over a rather small sampling regime, and only about 0.22% of insects were the biting flies targeted (Frick and Tallamy 1996). Extrapolations from this and an earlier survey implied that many billions of non-target insects could be killed.

References

Austin AD, Wharton RA, Dangerfield PC (1993) Revision of the endemic Australian subfamily Trachypetinae Schulz s.l. (including Cercobarconinae Tobias) (Hymenoptera: Braconidae). Invert Taxon 18:97–119

Borkent A, Brown BV, Adler PH, De Souza AD, Barber K et al (2018) Remarkable fly (Diptera) diversity in a patch of Costa Rican cloud forest: why inventory is a vital science. Zootaxa 4402: 53–90

Braby MF, Eastwood R, Murray N (2012) The subspecies concept in butterflies: has its application in taxonomy and conservation biology outlived its usefulness? Biol J Linn Soc 106:699–716

Brower AVZ (2006) Problems with DNA barcodes for species delimitation: 'ten species' of *Astraptes fulgerator* reassessed (Lepidoptera: Hesperiidae). Syst Biodivers 4:127–132

Brower AVZ (2010) Alleviating the taxonomic impediment of DNA barcoding and setting a bad precedent: names for ten species of '*Astraptes fulgerator*' (Lepidoptera: Hesperiidae: Eudaminae) with DNA-based diagnoses. Syst Biodivers 8:485–491

Cardoso P, Erwin TL, Borges PAV, New TR (2011) The seven impediments in invertebrate conservation and how to overcome them. Biol Conserv 144:2647–2655

Dennis RLH (1977) The British butterflies: their origin and establishment. EW Classey, Faringdon

Ellis S, Bourn N, Bulman C (2012) Landscape-scale conservation for butterflies and moths. Lessons from the UK. Butterfly Conservation, Wareham, Dorset

Fattorini S, Di Giulio A, Dapporto L (2013) Measuring insect rarity: practical issues, pragmatic approaches. J Insect Biodiv 1(10):1–21

Frick TB, Tallamy DW (1996) Density and diversity of nontarget insects killed by suburban electric insect traps. Entomol News 107:77–82

Harvey JA, Heinen R, Armbrecht I, Basset Y, Baxter-Gilbert JH et al (2020) International scientists formulate a roadmap for insect conservation and recovery. Nat Ecol Evol 4:174–176

Hebert PDN, Penton EH, Burns JM, Janzen DH, Hallwachs W (2004) Ten species in one: DNA barcoding reveals cryptic species in the neotropical skipper butterfly *Astraptes fulgerator*. Proc Natl Acad Sci U S A 101:14812–14817

Hribar LJ (2020) Expanding basic entomological knowledge by using mosquito surveillance bycatch. Ann Entomol Soc Am 113:439–446

Janzen DH, Hallwachs W (2016) DNA barcoding the Lepidoptera inventory of a large complex tropical conserved wildland, Area de Conservacion Guanacaste, norhwestern Costa Rica. Genome 59:641–660

Jin M, Zwick A, Slipinski A, Dekeyzer R, Pang H (2020) Museomics reveals extensive cryptic diversity of Australian prionine longhorn beetles with implications for their classification and conservation. Syst Entomol 45:745–770

Lawton JH, Brotherton PNM, Brown VK, Elphick C, Fitter AH et al. (2010) Making space for nature: a review of England's wildlife sites and ecological network. Report to DEFRA. Department of Environment, Food and Rural Affairs, London

Martoni F, Taylor GS, Blacket MJ (2020) Illuminating insights into the biodiversity of the Australian psyllids (Hemiptera: Psylloidea) collected using light trapping. Insects 11:354. https://doi.org/10.3390/insects11060354
New TR (1999) The evolution and characteristics of the Australian butterfly fauna. In: Kitching R, Scheermeyer E, Jones RE, Pierce NE (eds) Biology of Australian butterflies. CSIRO Publishing, Melbourne, pp 33–52
New TR (2007) Broadening benefits to insects from wider conservation agendas. In: Stewart AJA, New TR, Lewis OT (eds) Insect conservation biology. CABI, Wallingford, pp 301–321
Quicke DJL, Austin AD, Fagan-Jeffries EP, Hebert PDN, Butcher BA (2020) Recognition of the Trachypetidae stat. n. as a new extant family of Ichneumonoidea (Hymenoptera), based on molecular and morphological evidence. Syst Entomol 45:771–782
Rabinowitz D (1981) Seven forms of rarity. In: Synge H (ed) The biological aspects of rare plant conservation. Wiley, New York, pp 205–217
Raposo MA, Kirwan GM, Lourenco ACC, Sobral G, Bockmann FA, Stopiglia R (2020) On the notions of taxonomic 'impediment', 'gap', 'inflation' and 'anarchy', and their effects on the field of conservation. Syst Biodivers 19:296–311
Samways MA, McGeoch MA, New TR (2010) Insect conservation. A handbook of approaches and methods. Oxford University Press, Oxford
Smith MA, Rodriguez JJ, Whitfied JB, Deans AR, Janzen DH, Hallwachs W, Hebert PDN (2008) Extreme diversity of tropical parasitoid wasps exposed by iterative integration of natural history, DNA barcoding, morphology and collections. Proc Natl Acad Sci U S A 105:12359–12364
Stoechle MY, Hebert PDN (2008) Barcode of life. Sci Am 299(82–86):88
Tobias JA, Seddon N, Spottiswoode CN, Pilgrim JD, Fishpool LDC, Collar NJ (2010) Quantitative criteria for species delimitation. Ibis 152:724–746

Chapter 5
Drivers of Decline

5.1 Introduction: Categorising Threats

The variety of causes of insect declines and losses are almost impossible to evaluate fully, but the major categories below encompass many of the threats implicated most frequently—all, in essence, affect the places where insects live and their suitability to sustain insect populations and communities, or affect the insects directly. However, as Taylor et al. (2018) put it 'Most if not all key threatening processes affecting terrestrial or freshwater environments have a profound effect on insect and allied invertebrate populations'. Many of those key threatening processes of greatest significance are listed under various Australian legislations (p. 194) and most have far wider relevance than to insects alone.

Global environmental changes are rendered ever more complex by human activities, and many attempts have been made to categorise these stressors succinctly, recognising both the variety of drivers of change and that these interact in many intricate—and often unpredictable—ways. As Yang et al. (2019) summarised it 'Considered seperately, each driver of change has multiple manifestations. Their combined effects and interactions reflect the complexity of global change'. That complexity is linked with unpredictable outcomes, reflecting scale, extent and intensity of any combination of drivers, as 'threats'. Very broadly they encompass (1) changes in land use, leading directly to loss or change of ecosystems; (2) changes in the extent, variety and usage of chemicals, either directly or as more casual discards or pollution to both terrestrial and aquatic environments, and with numerous poorly documented sublethal and synergistic effects; (3) increased variety and extent of alien species, introduced deliberately or accidentally and affecting all environments they occupy or invade; and (4) climate changes, notably warming and increased precipitation and extreme 'weather event' frequency, affecting all environments and communities. Collectively, these drivers continue to influence insect communities through both direct impacts on survival, development, physiology and behaviour, and an array of changed resources and changed interactions between

© The Author(s), under exclusive license to Springer Nature Switzerland AG 2022

T. R. New, *Insect Diversity, Declines and Conservation in Australia*, Fascinating Life Sciences, https://doi.org/10.1007/978-3-030-90134-9_5

species (Moir et al. 2014). Predicting the impacts of multiple stressors on insects involves assessing synergistic and antagonistic effects rather than simply additive effects, and becomes an extremely difficult and uncertain exercise. Most studies on responses have involved only single threats, with far fewer attempts to appraise two stressors and larger numbers very rarely studied (Kaunisto et al. 2016). The sheer number of insect species and their possible interactions with 'thousands of stressors' (Kaunisto et al. 2016) thwarts any general predictive outcome beyond relative generalities. Complex interaction networks are modified continually as anthropogenic impacts and more natural conditions change, with effects mostly uneven on the partner taxa involved, and lead to both specific changes and wider alterations to food webs as balances across trophic levels also change (Murphy et al. 2020). The complexity noted here also means that widely used terms such as 'habitat loss', 'deforestation', 'urbanisation', 'agricultural intensification' and many others are deceptively simple without further qualification. Nevertheless, a global survey of opinion (Milicic et al. 2021) endorsed that the three major drivers of insect declines were the broad categories of agriculture, climate change and pollution, and that land protection and land management are predominant major needs for conservation and for mitigation of threats.

Many changes to natural environments affect their resident biota, but a few broad categories of anthropogenic impacts are of paramount concern, and are superimposed on more natural processes such as succession, and potentially modified by the impacts of climate changes. Most notably, habitat loss or degradation such as by clearing of natural vegetation to enable agricultural, urban and industrial development, changes to freshwater bodies such as by damming, channelling, drainage and pollution from human wastes and agricultural chemicals, and the general increased isolation of remaining natural or unspoiled sites across the largely altered landscapes, is a universal concern. Their extent and relative importance are complex to evaluate—about 75% of native vegetation has been lost in the arable areas of Australia (Braby et al. 2021), and the remainder is fragmented and disturbed by invasive species, other anthropogenic influences and susceptible to impact from climate changes.

Urbanisation is a key interest in insect conservation, with about six drivers contributing to species extinctions: Hahs et al. (2009) designated six for plants as (1) historical habitat transformation, (2) current land use, (3) environmental context, (4) urban environment; (5) human actions, and (6) biotic interaction. All have wider ramifications. Urban expansion continues, with threats to many environments near the perimeters of existing settlements as new suburbs and industrial centres are created, with sealed concrete or tarmac surfaces replacing vegetation. Many of Australia's major urban settlements fall within Hahs et al.'s (2009) "Type III" grouping, in which the built areas previously supported natural vegetation rather than being on already transformed land, so that insect habitats such as grassland, shrubland or forest are destroyed directly (New 2018). Their insect communities may differ considerably from those on urban areas with a longer history of more gradual landscape transformations. Many European cities, for example, are on lands transformed several centuries ago, from either natural vegetation or earlier agricultural uses. Such drastic changes to natural habitats are conspicuous, but many of the changes they induce are far more difficult to clarify or confirm. In and around towns and cities, for example, insect mortality from traffic impacts can be considerable—with some countered by recommendations for reduced speed limits at particular times in the flight season of especially significant species (New 2015, 2018). That mortality has only rarely been evaluated systematically. Daily collections of dead insects on the surface and sides of a 2 km length of highway in Ontario, Canada, over 4 months in each of 2012 and 2013 included 15 insect orders with a total of 117,675 individuals (Baxter-Gilbert et al. 2015). Corpses were dominated by Diptera, Hymenoptera and Lepidoptera, predominantly assessed as pollinators, but extrapolation from daily mortality rates to larger scales implied that vast numbers of insect deaths occurred. A more widespread change has been the increase of artificial lighting at night (recently dubbed 'ALAN'), which is claimed to present a growing threat to biodiversity, especially to nocturnal insects (Stewart 2021). Four main disruptive effects occur, reflecting that particular insect species or sexes may (1) be attracted to light, (2) actively avoid light, (3) have activity patterns moulded by particular photoperiods or (4) respond only to particular spectral patterns (RCEP 2009), but the most frequently expressed concern is concentrating phototactic moths at street and building lights, rendering them susceptible to increased predation or death through exhaustion if they are unable to break away from the light. Artificial lighting is commonly claimed to be a contributor to nocturnal insect declines, and its progressive increase into hitherto remote areas has gained considerable recent attention (Janzen and Hallwachs 2019) with recommendations to assess impacts suggested (Kalinkat et al. 2021). ALAN has been described as a 'potent ecological trap' for insects, and a feature to which insects have had no evolutionary opportunity to adapt and for which it can become a 'fatal attraction' (Owens et al. 2020). One account has estimated that 100 billion (10^{11}) insects perish at street lights each year in Germany, based on an estimated 8.2 million street lights and a third of the average 450 insects/night attracted to each being lost (Eisenbeis and Hanel 2009). Any such enormous 'vacuum cleaner effect' has clear potential to diminish local insect populations, and the worrying accounts of reduced insect numbers in light trap catches over recent years endorse the likelihood that ALAN is indeed a significant contributor to insect declines. Declines of moths and other nocturnal insects from street lights and related sources have wider ecological ramifications, such as influencing pollination. Examination of the probosces of light-captured moths in southern England, using swabs of fuchsin jelly for microscopical preparations, revealed that 23% of specimens examined carried pollen, collectively of 28 flower species and implying that moth removal at lights could reduce pollen transport (Macgregor et al. 2017).

ALAN is no longer purely an urban syndrome, with proliferation of light sources having impacts over much wider areas. Discussed by Owens et al. (2020), it is a complex array of threats because of the diversity of lighting structures, with some light sources accompanied by heat and/or ultrasonic emissions, and necessitating detailed awareness of insect behaviour if impacts are to be mitigated. Thus, it may be feasible to engineer LED sources for wavelengths that least influence insect fitness, as well as to limit illumination such as by dimming light sources to lower threshold acceptance levels.

Urbanisation has other subtle impacts on insects. Local changes in temperature can substantially influence insect environments. The 'urban heat island' effect reflects the effects of large areas of impervious surfaces that cause cities and towns to have warmer and drier local climates than nearby rural areas—with city temperatures reported as commonly 2–4 °C higher, affecting development and well-being of resident insects. Impacts on urban pollinators have attracted considerable attention (Ayers and Rehan 2021), but different species show different tolerances to warming and/or desiccation. Different responses to warming may influence the adequacy of interactions in urban areas as species' ranges increase or decrease or change in overlap, and in seasonal synchrony. Although the urban heat island effect is rarely considered specifically in this context, synchrony between urban plants and their pollinators may be disrupted: in Lille, northern France, high urbanisation was associated with earlier flowering periods but pollinator phenology was not affected (Fisogni et al. 2020).

High thermal tolerances may benefit both resident and invasive species. More generally, it is recognised widely that influences of climate change on insect ecology, development and behaviour are varied but also hard to distinguish clearly from the impacts of habitat loss and degradation, among the variety of other stressors to populations (Forister et al. 2021). For example, the relative influences of land use (loss of seminatural habitats to agriculture leading to losses) and global warming (enabling range extensions) on European grasshoppers appear to have changed over recent decades (Poniatowski et al. 2020).

The more subtle impacts of many human activities on insects are not yet documented fully. Many are undoubtedly overlooked or ignored widely. Inclusion of 'overexploitation' as threat to invertebrates (Cardoso et al. 2020), for example, is acknowledged by some rule of 'prohibition of take' in much legislation, but is difficult to evaluate, monitor or control. 'Overcollecting' of rare butterflies, beetles or dragonflies that can command high sale prices is the most common context (p. 157).

Any 'disturbance', potentially or actually a threat to insects, may obviously destroy the occupied site or critical resources needed or erode requirements in more subtle ways. A disturbance may (1) 'filter' the existing community by changing the pattern of resources available to benefit some species and deprive others and (2) facilitate changes in composition by admitting additional taxa that benefit from the changes but may adversely affect resident taxa through novel interactions. In some contexts, the nature of such changes may be anticipated based on ecological and functional traits of the taxa of interest. As one example, urbanisation effects on pollinators (above) have been appraised in many parts of the world (Ayers and Rehan 2021), leading to recognising the trends summarised in Fig. 5.1, some of which may apply more generally to a range of other taxa in similar environments. Again referring primarily to pollinators, LeBuhn and Vargas (2021) considered a series of abiotic and biotic 'filters' that influence the transformation of potential to actual species pool, noting that the influence of each reflects the individual context of

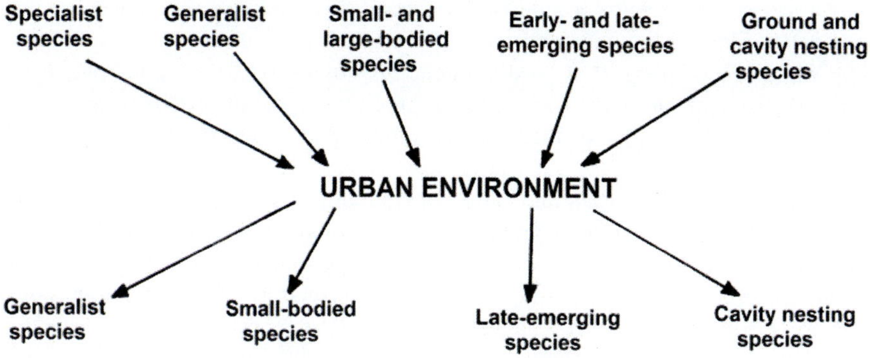

Fig. 5.1 Simplistic diagram of how some functional traits of natural bee communities are filtered to produce urban bee communities; the central 'net' represents a generic urban environment with the associated features that may influence bee community structure (based on diagram by Ayres and Rehan 2021; used under Creative Commons Attribution [CCBY] licence: https://creativecommons.org/licences/by/4.0/)

land use and management. Pollinator declines largely comprise specialist species that rely on increasingly vulnerable plant species in agricultural or urban landscapes, thereby declining from land use changes and exacerbated by abiotic changes and direct loss of resources. The 'abiotic filters' listed by LeBuhn and Vargas (impermeable surfaces, temperature increase, drought, pesticides, heavy metals) and 'biotic filters' (host plants, pathogens, introduced predators, introduced competitors) collectively cause severe alterations to community composition and health—and all may modify pollinator wellbeing directly but interact to increase unpredictability of influences. The 'urban filter' influences insect communities by selecting those species that can colonise and survive the harsh conditions of urbanisation because of their biological traits, many of which have functional implications, linked with loss of other insect diversity. It has five main contributory drivers (New 2015, and reviewed by Fenoglio et al., 2021): impervious surfaces (direct removal of habitat by cover of concrete, tarmac or construction), habitat fragmentation, urban heat island effect (with strong physiological impacts on many insects), pollution (light, noise, air and water), and alien plant species. These drivers can interact in many ways, mostly poorly understood.

Such analyses extend beyond individual species to consider the broad ecological features that may confer preferment in changed environments and that may collectively lead to more homogenous communities from which some ecological traits have been largely lost. 'Biotic homogenisation' from urbanisation has frequently been implied, but studies to clarify this have focused more on birds and plants than on arthropods (Knop 2016). Comparison of the richness of three groups of insects (Heteroptera, Auchenorrhyncha, Coleoptera) on birch trees in city and rural sites in Switzerland showed considerable differences, with both hemipteran groups revealing large-scale homogenisation through losses of specialist species from the city sites. Coleoptera gave more varied results, leading Knop (2016) to infer that cities might in some cases be more suitable for generalist species than rural areas and that urbanisation may have wide and complex impacts on local insect commmunities. One more general trend, described in numerous studies, is of 'gradients' of species richness from urban centres to adjacent rural environments. Richness generally declines toward the urban centres, largely reflecting the intensity of habitat change, but more perceptive ecological analyses, as above, are scarce. In temperate and tropical regions alike, urban and rural communities of insects differ substantially and, as well as the common gradient pattern, community composition depends strongly on the structure and variety of vegetation present. Comparisons of the geometrid moths across a gradient of land-use categories (forest, forest-pasture, pasture, urban) in Ecuador revealed strong differences, with urban assemblages far different from others and leading Gaona et al. (2021) to recommend larger open green urban spaces with high diversity of native plants: both these recommendations are frequent in urban insect conservation. In the above land-use sequence, numbers of geometrid species (and individuals) were: 218 (1442), 146 (840), 84 (406), 75 (376). More generally, increased floral diversity within urban areas has greatest potential benefit for ecologically generalised species, whose abundance can decrease substantially with increased urbanisation. The principle applies to constructs such as 'butterfly gardens' and 'dragonfly ponds', both immensely valuable assets in towns and cities.

The side-effects and synergies between the various major impacts on insect environments are often impossible to consider. In addition, and also likely to be altered in unpredictable ways as species' phenology and distributions change and community structures become distorted by climate change, the increased presences of alien species of plants and animals influence many long-coadapted interactions between native species of insects, hostplants, and more general food web structures. Competition for such resources between susceptible indigenous ecologically specialised insects and aggressive generalist alien taxa is widespread, and a major cause of insect losses in isolated environments. Clarifying the incidence and relative importance of these drivers (often 'stressors' or 'threats') and understanding their impacts on both individual insect species and wider insect abundance and diversity is the core foundation for conservation action. The balance of local threats among the above broad categories may help to indicate local or national priorities. For Britain, Goulson (2019) noted the main categories of habitat loss, exposure to 'complex mixtures of pesticides' and the beginnings of climate change impacts. All are major contributors to insect vulnerability in Australia, under guises such as vegetation clearing for urbanisation, industrialisation and recreation, agricultural development, changes to water bodies, and are confounded by impacts from a large array of alien species, many of them widespread. Agricultural intensification is a leading driver of many reported insect declines in the northern hemisphere, especially from the middle decades of the last century, when trends such as removal of field boundaries (such as hedgerows), clearing of forests (most significantly in the tropics) and changes of local wetlands to create larger cultivable units, increased use of pesticides and fertilisers, and adoption of monocultures over large areas led to marked reductions in habitat heterogeneity and diversity on landscape scales. For southern Australia, land clearing, effects of sheep and cattle stocking and the impacts of other feral animals are the most significant threats to trapdoor spiders (Rix et al. 2017) and, undoubtedly, also to many other ground-dwelling invertebrates. As on an 'island continent', Australian environments may be also especially vulnerable to invasive alien species as these invade complex communities with numerous specialised species; weed and pest arthropod invasions of small vegetation remnants and reserves are a continuing concern.

Whilst different stressors can interact and compound or confuse predictions of their impacts, two spatial scales of stressors are sometimes distinguished. Thus, Wagner et al. (2021) recognised (1) large (in their discussion, continental) scale stressors including reduced habitat quality and extent from land-use changes and climate changes, with lesser influences from introduced species and atmospheric nitrification, and (2) site-specific threats, including pesticide use and light pollution. Atmospheric nitrogen levels, rarely considered specifically in insect conservation, are claimed to have already caused considerable changes in assemblages and community structure: Habel et al. (2016) noted habitat generalists becoming more dominant, and fewer species with specific habitat demands—including species adapted to habitats with poor nutrient loads, and monophagous taxa. In conjunction with climate changes, nitrogen loads can strongly influence interactions between species. Attempts have been made to stabilise definitions of threats to biodiversity in order to facilitate standard use in planning conservation management (Salafsky et al. 2008), and there is general agreement that assessing, defining and some ranking of the threats present for severity and urgency of remediation, and how these may be mitigated or prevented is the core of many conservation plans and activities.

Near-immediate losses (extirpations) of insects from sites (or, more rarely, from wider landscapes) from habitat change are the most usual context for conservation interest. However, even extensive habitat loss may not cause rapid losses but, rather, lead to extinctions that may not occur for some time after the disturbance and in part reflect the new equilibrium of that community. They are often unnoticed or attributed to the historical cause but, reflecting the large proportion of some natural habitats lost or degraded over recent decades, some workers have suggested that extinction debt may be common in many natural communities (Kuussaari et al. 2009). The number or proportion of specialist species (or the number or proportion of population units in metapopulations) expected to become locally extinct constitute the extinction debt. Essentially, without any further disturbances, many such species will be lost—but, even when this is not recognised, the current presence of those species implies chances of conservation. Extinction debt is difficult to detect, but rapid responses to habitat degradation by restoration or other management may effectively counter losses. For European calcareous grasslands, that management includes maintaining large plots by traditional management practices (mowing once a year or rough grazing) and restoration to promote connectivity within the severely fragmented landscape (Loffler et al. 2020). The seasonal timing of management may also be important.

5.2 Understanding Habitat Change

'Habitat change', universal but a concept sufficiently broad to need qualification in any given context, is over-riding and embraces many of the themes often treated as separate threats or 'drivers'. Thus, Edge and Mecenero (2015) separated the broad categories of 'Habitat degradation from anthropogenic alteration of ecosystem functions' and 'Habitat destruction through irreversible landscape transformation'. Each includes several key themes (Table 5.1) that may each have prevalent relevance in any individual context, and each was evident and exemplified for particular South African butterflies across varied contexts. Thus, 'Alteration of herbivore roles in the ecosystem' can be due to introduced animals (accidentally or deliberately) and loss of historically important megaherbivores, leading respectively to direct reductions of host plants by herbivory, transformed vegetation structure and microclimate, and in turn leading to losses of host plants (or the host ants of some Lycaenidae). Three main trends of habitat change are important for insect conservation, as (1) direct habitat loss, (2) decrease in habitat quality that affects occupancy rates and carrying capacity for richness of and/or biomass, and (3) increased isolation of remaining suitable or unthreatened habitat patches, changing patterns of gene flow and metapopulation dynamics.

Understanding the term 'habitat', as perhaps the most frequently cited axis of loss, is central to evaluation of potential threats to insects and communicating those concerns effectively. It implies much more than the most traditional definition as simply 'a place to live'. Many entomologists define habitat for terrestrial insects in terms of vegetation structure, with parallels to 'biotope', but recent changes in the concept have been a stimulus to more careful thought in insect conservation.

Table 5.1 Key extrinsic threats to butterflies in South Africa, with major subcomponents implicated for threatened taxa (from Edge and Mecenero 2015)

Main threat	Sub-threats
Habitat degradation from anthropogenic alteration of ecosystem functions	
	Introduced invasive organisms
	Inappropriate fire regimes
	Alteration of herbivore roles in ecosystems
Habitat destruction through irreversible landscape transformation	
	Agriculture and forestry
	Urbanisation
	Mining and industrial developments
Anthropogenically induced environmental changes	
	Global warming from increased greenhouse gas emissions
	Industrial pollution
	Agricultural chemicals
Direct killing of the organism	
	Removal of live material from the wild

Following ideas developed by Dennis et al. (2006) and Dennis (2010), that 'place' is, rather, an environment in which a species' critical resource needs occur in sufficient supply and are accessible to the species that depend/s on them. Those resources comprise two main categories: (1) 'consumables', specific food materials needed to support all feeding stages, such as larval foodplants or prey, and adult supplies of nectar, pollen or prey, and any mutualist or symbiotic taxa that are obligatory associates of the consumer; and (2) 'utilities', the environmental features that enable the species to undertake its full repertoire of normal development and behaviour—such as suitable pupation, overwintering or aestivation sites, basking and patrolling areas, territorial posts, nesting sites, areas free from natural enemies, and so on. Gaining these key resources may then be affected by 'conditioners', such as suitable climate and disturbance regimes influencing insect activity and physical access.

Each species will have a somewhat different suite of critical resources, and these may change over a species' range—many insect herbivores primarily use different food plant species across their range, and different species of mutualistic ants may associate with lycaenid butterflies, as examples. Patterns of resource availability are not necessarily static; they may change naturally (as by vegetation succession), seasonally, from disturbance or interactions with other species, so that plans to sustain these on a given site may require a wider perspective. An alien plant species invading a sensitive area, for example, may be adopted as an important food plant by a native insect herbivore and—as for the Golden sun-moth on Australian native grasslands (p. 200)—itself become a critical resource for a threatened species.

5.3 Understanding Threats

The above encapsulates the four major 'drivers of decline' listed in presumed order of importance by Sanchez-Bayo and Wyckhuys (2019) as reflected in the number of attributions in the studies they reviewed, namely (1) habitat loss and conversion by intensive agriculture and urbanisation; (2) pollution, mainly by synthetic pesticides and fertilisers; (3) biological factors, including pathogens and introduced species; and (4) climate change. Each is very complex, as noted above, and some form of their impacts (past, current or anticipated) is the basis of most practical insect conservation need. Impacts affect either or both of the supply of critical resources or the direct survival and 'performance' of the insects themselves. Mortality is a common outcome from pollution or disease, for example, but sublethal impacts such as slower growth rates, smaller size, behavioural changes and decreased reproductive capacity are usually unknown. Likewise, the changed ecological interactions flowing from communities changed by loss of residents and/or invasions by or presence of alien species are often very subtle and evident only from the hindsight of changes once they have occurred. Many of the threats nominated also have synergistic effects, which may be additive or more complex. Climate change, for example, may differentially affect activity and wellbeing of parasitoids and predators and their insect hosts or prey, so that intensities of attack and levels of mortality are changed. Resulting changes to seasonal appearance or development can disrupt species interactions, but impacts on ecological functions and population dynamics may be far more subtle and their effects on the wider community hard to predict or measure. As another example, impacts on conditions needed by nesting bees (many of which have highly specific nesting requirements) include changed exposure and amount of bare ground, and the availability of above-ground cavities such as plant stems (Harmon-Threatt 2020). In turn, these changes affect bees' ability to found nests and their later growth, development and overwintering success—the last because larvae and other overwintering stages have only minimal capability to respond to threats.

Most of these threats have very widespread impacts, by no means only on insects. Agricultural intensification (including increased pesticide uses) and light pollution (especially in urban areas), as examples, are cited most frequently for their effects on insects rather than on other organisms that may be at least equally affected, or for their wider relevance to diversity and community composition. However, many threats are in principle reversible by focused management—so that agricultural practices can be modified, and light regimes and intensity changed. In contrast, others (notably the impacts of alien species and climate change) are essentially irreversible. Changed interactions from invasive species are often regarded as 'permanent' (Wagner and Van Driesche 2010) and far-reaching across multiple species in affected regions. Reviewing the relative threat categories as applied to the 57 endangered/threatened insect species then listed for the United States, Wagner and Van Driesche ranked each of the 17 threat categories for each species as (1) primary threats, those mentioned specifically in published sources as being of immediate concern; (2) secondary threats, those of lesser concern and not of immediate or chronic importance; and (3) tertiary threats, largely implied or potentially problematic but lacking evidence of any impact. 'Invasive species' were implied to be one of the most serious threats in all three categories, and affected more of the endemic listed species (42) than any others. In a second group of species reviewed by Wagner and Van Driesche (2010), invasive species as secondary and tertiary threat components were also dominant for the rare Lepidoptera of woodlands in eastern North America.

However, one strong bias in that survey of listed species was that invasive species were a primary level threat to all 13 listed species from Hawaii, where invasive species have for long been a major concern for that remarkable isolated endemic biota (Howarth 1991, 2001), but for only a small proportion (11%) of the U.S. mainland taxa. In general, fauna of long-isolated oceanic islands, with substantial independent evolutionary history may be highly susceptible to aggressive or generalist alien species, with the endemic taxa especially vulnerable to such imposition.

Threat evaluation is central to all aspects of insect conservation, and all the major threats have arisen repeatedly and have been discussed extensively in recent accounts (Samways 2020). Their variety and influences in very different contexts have engendered numerous counter-measures, remedial approaches, means of tolerance or avoidance at the core of conservation management programmes, in which phrases such as 'threat abatement' and 'threat avoidance' are integral directors of activity.

Many action plans for insect conservation, whether for single species or wider targets, are largely based on that premise in combination with habitat protection—if necessary including restoration and remediation measures—and promoting connectivity (that is, countering or avoiding isolation) through habitat networks, corridors and other landscape modifications that help to promote an insect's normal population processes. In conjunction, education and public awareness of threats is a fundamental need: as has been emphasised frequently, threat containment and control demand cooperation between all facets of society. Some such measures are accepted widely as desirable—reductions in pesticide use, 'cleaning' of polluted waters, protection of key biotopes such as tropical forests, replacement of intensive agriculture by less environmentally damaging procedures, and biosecurity measures to reduce importations of alien species in commerce or passenger luggage are all components of management that are received sympathetically, and supported widely.

The major advances in recognising and alleviating threats to insects, developed mainly in the northern hemisphere, form a basis for emulation wherever insect conservation is needed. However, the far different social and scientific backdrops over much of the rest of the world commonly render uncritical transfer of even the most effective and desirable conservation strategies for insects very difficult to achieve. Thus, insects in the southern temperate regions have an equivalent variety of threats to insect life to those in the northern hemisphere, but capability to address these is very limited, not least because detailed knowledge of the fauna is relatively embryonic (New and Samways 2014). In contrast to the better-known insect faunas of much of the northern hemisphere, insects in 'the south' play far smaller roles in conservation agendas—and, reflecting paucity of taxonomic and ecological knowledge, their conservation needs are often more difficult to study and evaluate, and subsequently to define and support continuing management over extended periods.

Related concerns are perhaps even greater elsewhere—forest clearance and chemical uses over vast and highly under-documented areas of the tropics, where formal attention to insect conservation is in its infancy or non-existent, political and practical problems of human welfare dominate national concerns, and economic reward from exploitation of natural resources is a mainstay of continued survival leave little flexibility for those transfers. Those areas also harbour massive insect richness and diversity presumed to be threatened by despoliation of their habitats.

Despite clear evidence of direct habitat losses and other threats in many such environments, resources for the 'luxury pursuit' of insect conservation are simply not available, and any chances of acceptance of work on insects beyond the direct economic or welfare aspects of managing key crop pests and disease vectors are usually slight. Even when resident entomologists are employed, their priorities and responsibilities inevitably devolve largely on insects of economic importance, mostly in crop protection and human welfare. Any conservation needs of insects in arenas lacking such tangible support are likely to be met 'passively', perhaps fortuitously through occurring in areas reserved to protect charismatic mammals and birds as foci for ecotourism, or in cultural areas such as totem forests, rather than addressing threats directly. Rare cases of more constructive and deliberate insect conservation in the tropics can be fostered through economic benefit. The establishment of the Insect Farming and Trading Agency (IFTA) in Papua New Guinea was a notable example which focused on ranching of endemic butterflies (most notably many of the spectacular 'birdwings', *Ornithoptera* spp., Papilionidae) desired by collectors throughout the world (Parsons 1992). A network of, at its peak, several hundred people scattered throughout the country undertook this ranching, which involved establishing and increasing numbers of the larval food plant vines in gardens (as habitat enrichment), increasing their availability to—and attracting—female birdwings to oviposit. Larvae were then reared to adulthood, protecting them from losses by predators and parasitoids, and a proportion of the emerging adults retained for sale, with the rest liberated. Sales were coordinated by the Agency, with CITEs permits and any other necessary export documents, and collectors were thus able to purchase 'perfect specimens' of highly desirable species, and a wide variety of other insects collected by the participants. Birdwings, including the world's largest butterflies, are a prime local flagship group for the protection of tropical forest—especially on lowland areas, much has been cleared for establishment of oil palm plantations (Mitchell et al. 2016), and the need to protect that forest as butterfly habitat, and a source of income for local people has become more widely accepted. Augmenting small reward from subsistence farming, butterfly ranching enabled some communities to participate in a cash economy for the first time in their history (Parsons 1992).

As in some other exercises in tropical insect conservation, IFTA was founded from the concerns and initiative of expatriate entomologists. However, simplistically, some dictates from entomologists versed in scenarios of the northern temperate regions and lacking practical experience of the difficulties and social priorities elsewhere may not transfer easily; direct transfer of even very successful approaches may simply not be possible. In short, anthropogenic pressures on the habitats of tropical and southern region insects parallel those on northern temperate fauna. They are, though, far more difficult to address, from both motivation and capability (Samways 1995, New and Samways 2014).

5.4 A Geographical Pattern?

Experience and knowledge of the world's insects and their conservation needs demonstrate a clear geographical gradient, with three major distinct arenas. The northern temperate zone is by far the most thoroughly documented and, despite remaining gaps, is the crucible of conservation concern and practice. The foundations of awareness and practice, especially from concerns for butterflies in Britain, western continental Europe and North America, have provided the rationales and protocols for emulation elsewhere, and much of the understanding, philosophy and practice of modern insect conservation derives directly from the numerous successes or failures in that region. Second, the insects of the southern temperate zone, with much in common with the northern zone, are far more poorly known, but with sufficient to confirm that conservation needs are indeed substantial and major threats definable and able to be considered constructively. Conservation experiences and approaches can broadly parallel those from the northern regions, with due allowances for logistic and information limitations, and the far larger number of insect species present restricting reliance on the 'single species' level of conservation.

Last, the vast and complex insect faunas of the intervening tropical zones are far more poorly known, and practical conservation devolves largely on conservation of habitats, most notably key vegetation types such as forests, that continue to be under immense pressures from destruction and change. Protected areas, maintained free of human interventions, are the major practical option for insect conservation. Over much of the tropics, single insect species conservation is unusual, many of the lessons from elsewhere are impracticable and local concerns for insect wellbeing and its advocates are often minimally heeded 'voices in the wilderness' with little high level local support. Progressive enumerations of tropical forest insects (Chap. 3), in particular, demonstrate the urgent need for conservation of those forests as harbours of almost untold richness and complexity, with calls for 'core areas' of unspoiled forest to be secured in perpetuity and buffered by resuming (wherever possible) the anthropogenic adjacent areas and allowing them to add to the effective reserve area.

Australia is unique in supporting extensive endemic and globally significant insect faunas characterising both tropical and temperate zones within the same national purview and legislation, and so potentially enabling conservation spanning the priorities and restrictions of approaches cited above for either individual zone. Complementary approaches and strategies, from individual species conservation to safeguarding regional and national diversity based on protection of key habitats and areas, with the benefits from studies and examples throughout the world setting bases for novel modifications as needed to assure the best possible outcomes. That idealistic scenario draws on the entomological perspective discussed in the next chapter.

References

Ayers AC, Rehan SM (2021) Supporting bees in cities: how bees are influenced by local and landscape features. Insects 12:128. https://doi.org/10.3390/insects12020128

Baxter-Gilbert JH, Riley JL, Neufeld CJH, Litzgus JD, Lesbarreres D (2015) Road mortality potentially responsible for billions of pollinating insect deaths annually. J Insect Conserv 19: 1029–1035

Braby MF, Yeates DK, Taylor GS (2021) Symposium overview. Population declines and the conservation of insects and other terrestrial invertebrates in Australia. Austral Entomol 60:3–8

Cardoso P, Barton PS, Birkhofer K, Chichorro F, Deacon C et al (2020) Scientists' warning to humanity on insect extinctions. Biol Conserv 242:108426

Dennis RLH (2010) A resource-based habitat view for conservation. Butterflies in the British landscape. Wiley-Blackwell, Oxford

Dennis RLH, Shreeve TG, Van Dyck H (2006) Habitats and resources: the need for a resource-based definition to conserve butterflies. Biodivers Conserv 15:1943–1966

Edge DA, Mecenero S (2015) Butterfly conservation in southern Africa. J Insect Conserv 19:325–339

Eisenbeis G, Hanel A (2009) Light pollution and the impact of artificial night lighting on insects. In: McDonnell MJ, Hahs AH, Breuste JH (eds) Ecology of cities and towns. Cambridge University Press, Cambridge, pp 243–263

Fenoglio MS, Calvino A, Gonzalez E, Salvo A, Videla M (2021) Urbanisation drivers and underlying mechanisms of terrestrial insect diversity loss in cities. Ecol Entomol 2021. https://doi.org/10.1111/een.13041

Fisogni A, Hautekeete N, Piquot Y, Brun M, Vanappelghem C et al (2020) Urbanization drives an early spring for plants but not for pollinators. Oikos 129:1681–1691

Forister ML, Halsch CA, Nice CC, Fordyce JA, Dilts TE et al (2021) Fewer butterflies seen by community scientists across the warming and drying landscapes of the American west. Science 371:1042–1045

Gaona FP, Iniguez-Armijos C, Brehm G, Feder K, Espinosa CI (2021) Drastic loss of insects (Lepidoptera: Geometridae) in urban landscapes in a tropical biodiversity hotspot. J Insect Conserv. https://doi.org/10.1007/s10841-021-00308-9

Goulson D (2019) The insect apocalypse, and why it matters. Curr Biol 29:R967–R971

Habel JC, Segerer A, Ulrich W, Torchyk O, Weisser WW, Schmitt T (2016) Butterfly community shifts over two centuries. Conserv Biol 30:754–762

Hahs AK, McDonnell NJ, McCarthy MA, Vesk PA, Corlett RT et al (2009) A global synthesis of plant extinction rates in urban areas. Ecol Lett 12:1165–1173

Harmon-Threatt AN (2020) Influence of nesting characteristics on health of wild bee communities. Annu Rev Entomol 65:39–56

Howarth FG (1991) Environmental impacts of classical biological control. Annu Rev Entomol 36: 485–509

Howarth FG (2001) Environmental issues concerning the introductions of non-indigenous biological control agents. In: Lockwood JA, Howarth FG, Purcell MF (eds) Balancing nature: assessing the impacts of importing non-native biological control agents (an international perspective). Entomological Society of America, Lanham, MA, pp 70–99

Janzen DH, Hallwachs W (2019) Perspective: where might be many tropical insects? Biol Conserv 233:102–108

Kalinkat G, Grubisic M, Jechow A, van Grunsven RHA, Schroer S, Holker F (2021) Assessing long-term effects of artificial light at night on insects: what is missing and how to get there. Insect Conserv Divers 14:260–270. https://doi.org/10.1111/icad.12482

Kaunisto S, Ferguson LV, Sinclair BJ (2016) Can we predict the effects of multiple stressors on insects in a changing climate? Curr Opin Insect Sci 17:55–61

Knop E (2016) Biotic homogenization of three insect groups due to urbanization. Glob Change Biol 22:228–236

Kuussaari M, Bommarco R, Heikkinen RK, Helm A, Krauss J et al (2009) Extinction debt: a challenge for biodiversity conservation. Trends Ecol Evol 24:5654–5571

LeBuhn G, Vargas J (2021) Pollinator decline: what do we know about the drivers of solitary bee declines? Curr Opin Insect Sci 46. https://doi.org/10.1016/j.cois.2021.05.004

Loffler F, Poniatowski D, Fartmann T (2020) Extinction debt across three taxa in well-connected calcareous grasslands. Biol Conserv 246:108588

Macgregor CJ, Evans DM, Fox R, Pocock MJO (2017) The dark side of street lighting: impacts on moths and evidence for the disruption of nocturnal pollen transport. Glob Change Biol 23:697–707

Milicic M, Popov S, Branco VV, Cardoso P (2021) Insect threats and conservation through the lens of global experts. Conserv Lett 2021:e12814. https://doi.org/10.1111/conl.12814

Mitchell DK, Dewhurst CF, Tennent J, Page WW (2016) Queen Alexandra's birdwing butterfly *Ornithoptera alexandrae* (Rothschild 1907): a review and conservation proposals. Southdene Sdn Bhd, Malaysia

Moir ML, Hughes L, Vesk PA, Leng MC (2014) Which host-dependent insects are most prone to coextinction under changed climates? Ecol Evol 4:1295–1312

Murphy SM, Richards LA, Wimp GM (2020) Editorial: arthropod interactions and responses to disturbance in a changing world. Front Ecol Evol 8:93. https://doi.org/10.3389/fevo.2020.00093

New TR (2015) Insect conservation and urban environments. Springer, Cham

New TR (2018) Promoting and developing insect conservation in Australia's urban environments. Austral Entomol 57:182–193

New TR, Samways MJ (2014) Insect conservation in the southern temperate zones: an overview. Austral Entomol 53:26–31

Owens ACS, Cochard P, Durrant J, Farnworth B, Perkin EK, Seymoure B (2020) Light pollution is a driver of insect declines. Biol Conserv 241:008259

Parsons MJ (1992) The world's largest butterfly endangered: the ecology, status and conservation of *Ornithoptera alexandrae* (Lepidoptera: Papilionidae). Tropical Lepidoptera 3(Suppl 1): 35–62

Poniatowski D, Beckmann C, Loffler F, Munsch T, Helbing F et al (2020) Relative impacts of land-use and climate change on grasshopper range shifts have changed over time. Glob Ecol Biogeogr 29:2190–2202

RCEP (Royal Commission on Environmental Pollution) (2009) Artificial light in the environment. Her Majesty's Stationery Office, London

Rix MG, Huey JA, Main BY, Waldock JM, Harrison SE et al (2017) Where have all the spiders gone? The decline of a poorly known invertebrate fauna in the agricultural and arid zones of southern Australia. Austral Entomol 56:14–22

Salafsky N, Salzer D, Stattersfield AJ, Hilton-Taylor C, Neugarten R et al (2008) A standard lexicon for biodiversity conservation: unified classifications of threats and actions. Conserv Biol 22: 897–911

Samways MJ (1995) Southern hemisphere insects: their variety and the environmental pressures upon them. In: Stork NE, Harrington R (eds) Insects in the changing environment. Academic, London, pp 297–320

Samways MJ (2020) Insect conservation. A global synthesis. CABI, Wallingford

Sanchez-Bayo F, Wyckhuys KAG (2019) Worldwide decline of the entomofauna: a review of its drivers. Biol Conserv 232:8–27

Stewart AJA (2021) Impacts of artificial lighting at night on insect conservation. Insect Conserv Divers 14:163–166. https://doi.org/10.1111/icad.12490

Taylor GS, Braby MF, Moir ML, Harvey MS, Sands DPA et al (2018) Strategic national approach for improving the conservation management of insects and allied invertebrates in Australia. Austral Entomol 57:124–149

Wagner DL, Van Driesche RG (2010) Threats posed to rare or endangered insects by invasions of nonnative species. Annu Rev Entomol 55:547–568

Wagner DL, Fox R, Salcido DM, Dyer LA (2021) A window to the world of global insect declines: moth biodiversity trends are complex and heterogeneous. Proc Nat Acad Sci 118(2): e2002549117

Yang LH, Postema EG, Hayes TE, MacArhur-Waltz DJ (2019) The complexity of global change and its effects on insects. Curr Opin Insect Sci. https://doi.org/10.1016/j.cois.2021.05.001

Chapter 6
Insect Conservation Need in the Southern Hemisphere

6.1 Insects of the Temperate Southern Hemisphere

A major collective difference—beyond the characteristics and study history of the insects themselves—between the northern and southern temperate zones becomes important in assessing insect faunas and their needs for conservation in the latter regions, namely their contrasting geographical and political environments. National interests and priorities among insect taxa in each country of mainland Europe, for example, overlap with the fauna of contiguous and other regional countries, because numerous insect species transcend political boundaries and occur in at least several countries with independent jurisdictions and conservation interests. They may have different conservation status and needs in each of these, as being threatened in only parts of a wider contiguous range. Likewise, Canada. and the United States share a long border in North America, and have many insects in common but with some northward attenuation of diversity as temperatures decrease. As Stewart and New (2007) noted, the southern temperate regions, collectively with far smaller land areas than the northern areas, are disjunct and long isolated from each other. The environments and biota of each of southern South America, southern Africa, Australia and New Zealand are thus largely independent in their recent evolutionary history. Australia and New Zealand have no land boundaries with any other countries; insect conservation interests in southern Africa are concentrated predominantly in South Africa, and in each of these three nations, independent national interests in insect conservation have made substantial strides. Relatively little parallel progress has been made in southern South America, and the insects of that region are still relatively poorly known (Contador et al. 2012). However, the very isolation of the Magellanic and sub-Antarctic regions of Chile and Argentina itself confers protection from some outside pressures other than, perhaps, some alien invasive species such as the generalist predatory social wasp *Vespula germanica*.

Each of Australia, New Zealand and South Africa has notable groups of flagship insects as foci for future expansion of insect conservation interests (New and

© The Author(s), under exclusive license to Springer Nature Switzerland AG 2022

T. R. New, *Insect Diversity, Declines and Conservation in Australia*, Fascinating Life Sciences, https://doi.org/10.1007/978-3-030-90134-9_6

Samways 2014). Most are of sufficient global interest, mainly as representatives of ancient isolated 'southern lineages' of insects that do not occur elsewhere, to reflect on wider policies and schedules. As one example, the endemic Australian Ancient greenling damselfly (*Hemiphlebia mirabilis,* p. 197) was previously accorded global priority by the Species Survival Commission's Odonata Specialist Group (Moore 1997), New Zealand's giant weta taxa (Orthoptera) penetrate many aspects of the country's culture, and South Africa's Elephant dung beetle (*Circellium bacchus*) is familiar to many tourists through appearing on road signs advising that 'Dung beetles have right-of-way'. All are iconic taxa of considerable importance in national and international advocacy. Their conservation is necessarily a melding of lessons from 'the north' and the individuality that reflects the local environments, capability and needs.

Across the four southern temperate regions, numerous insect taxa are related through their shared Gondwanan ancestry. Long isolated, both Australia and New Zealand have strong evidence of this heritage in their insect faunas, with their development moulded by other environmental idiosyncrasies: for example, New Zealand generally lacked native terrestrial mammals (other than bats), a trait facilitating development of large flightless ground-dwelling insects largely free from vetebrate predation. Weta, above, are a notable such case, with some larger species gaining the name of 'invertebrate mice'. One of the most severe current threats to New Zealand insects is predation by introduced mammals (Watts et al. 2012).

In both countries, the endemic insects are no less 'peculiar' and remarkable than the more charismatic vertebrate fauna to which far greater conservation attention has been given. As the predominant flagship group of insects for New Zealand, weta parallel the role of butterflies in most other places. Notwithstanding the occurrence of several species complexes of endemic butterflies in New Zealand (Patrick and Patrick 2012), it is indeed unusual for them to be subsumed in this way: for most other countries, butterflies are the paramount flagship group for fostering insect conservation interest. Butterflies are far richer in Australia (Braby 2000, 2016) and South Africa (Ball 2012) than in New Zealand, and accumulated information in large part reflects historical collector interests. National conservation overviews for butterflies of Australia (Sands and New 2002) and South Africa (Henning and Henning 1989) provided foundations for further progress in conservation assessments and biological study. More recent conservation overviews of South African butterflies (Mecenero et al. 2013, 2021) have set an enviable basis for conservation priorities, needs and actions. The latter account updated the conservation status assessments of the almost 800 taxa treated in 2013, and provided accounts of 165 threatened and priority taxa, with considerable new distributional information helping to refine determinations of conservation status. Formal attention to insect conservation has increased in all three countries, as summarised by New and Yen (2012, Australia), Watts et al. (2012, New Zealand) and Samways et al. (2012, South Africa). However, and despite considerable attention to other insect groups such as Odonata in South Africa (Samways 2008) and Australia (Theischinger and Hawking 2021), the unique national perspective provided for New Zealand by documenting all native threatened invertebrates as comprehensively as possible (McGuiness 2001) and preparing national recovery plans for weta and carabid beetles has enabled focused advances evaluated against realistic priorities and opportunities.

Against a formidable ignorance of the insects of these southern zones and the tropical areas to their north, it is easy to overlook that the biology of many insect species in the 'well-known' northern temperate regions is still unknown in detail, and thwarts attempts to include them fully in conservation planning. For the United States, insects 'still exist outside mainstream conservation policy research and implementation' (Bossart and Carlton 2002), a contention attributed in part to issues of credibility arising from 'an appalling lack of basic knowledge of many diverse and ecologically important insect taxa' that can weaken belief in the real conservation needs of species promoted for listing or other attention. The problems in 'listing' insects recognised by Bossart and Carlton reflected the 'prohibitive logistical challenges' that confront those who must take decisions involving numerous taxa for which little (if any) biological information is available—and this is recognised widely as a serious impediment to species listing for invertebrates in general (New 2009a, b, Pyle 2012). Pyle also discussed an often-overlooked consequence of 'listing' species in the United States, and paralleled elsewhere—that the act of formally listing a species for protection functionally prevents or threatens efforts to gain the very biological information needed for sensible fact-based conservation. He noted the more specific context in which concerned specialist workers provide the information on which the initial listing is justified but are then restricted from further study and, effectively, from participating in the conservation endeavour. Well-intentioned protection of an insect can thus prove counter-productive through imposing blanket bans on taking specimens or access to study sites, even by the people most capable of furthering understanding and optimising focus of conservation management. Deterring the small amounts of real support potentially available in this way can decrease support for legislative listing, and also lead concerned hobbyists and others not to contribute to such exercises; increased suspicions can weaken the progress from initial listing to informed conservation practice. When only very restricted informed support is likely to be available, as in Australia, it must be identified and encouraged as strongly as possible. In many parts of Australia, local entomological expertise is very limited, and a few local hobbyists may be the predominant holders of relevant knowledge that can contribute most to conservation needs. Responsible acquisition of voucher specimens to verify identity is necessary for many 'hard-to-identify' insect groups. Without these, reports of incidences of, for example, some lycaenid and hesperiid butterflies from sites outside their recorded range remain open to doubt. Even the high quality photographs now so readily obtained are not sufficient for validation of some taxa.

6.2 Australia's Insects and Conservation Need

Australia (1) contains a very varied range of terrestrial and inland water environments inhabited by endemic and other insects; (2) extends from tropical to cool temperate regimes and from humid to arid environments; (3) has a relatively short history of anthropogenic disturbance from European settlement superimposed on a far longer period of First Peoples' land management; and (4) a large insect fauna, with inadequate attention to its future wellbeing and the threats it faces. Australian insect variety has been appreciated since the early days of European exploration (New 2011), and the early history of its discovery and characterisation described in considerable detail (Marks 1991; Upton 1997). Many taxonomic groups have been monographed or revised over recent decades, and the synopses of all insect orders given in Naumann (1991) is a significant account of then contemporary knowledge. Surveys of Australia insects continue to reveal large numbers of species, but also taxonomic 'oddities' of global significance in aiding understanding of insect relationships, evolution and phylogenies as they are studied progressively by morphological and molecular techniques. The tiny moth *Aenigmatinea glatzella* from Kangaroo Island (South Australia), for example, constituted an entirely new family (Aenigmatineidae) of homoneurous moths that led to rearrangement of relationships within the Lepidoptera (Kristensen et al. 2015). Also unusually, its larvae feed as stem-miners in *Callitris* (Cupressaceae). As another example, when Rentz (1980) erected the new endemic orthopteran family Cooloolidae to accommodate the 'Cooloola monster' (*Cooloola propator*) he commented that his initial impression of the insect was of a 'manufactured species'—so paralleling early reactions to the Platypus when specimens were first seen in Europe. All species of Cooloolidae are very localised and rarely seen. However, one (the largest of the four described species, *Cooloola ziljan*) is known only from around a sugar cane field in Queensland, and Rentz (1994) feared for its future in that highly disturbed landscape as the farm was then in the process of being sold. He surmised that the other three species might be more secure, as all occurred in national parks.

In marked size contrast to the small *Aenigmatinea* moth, even Australia's largest insects can be elusive. One of the largest moths, the Atlas moth (*Attacus wardi*, Saturniidae) was not rediscovered for many years after it was described (p. 183). A more recent discovery, the Gargantuan stick insect (*Ctenomorpha gargantua*) is one of the world's largest insects; body length of the male type specimen was 189 mm. Described only in 2006, and known from few specimens collected in the tropical forests of north Queensland, this gigantic insect is apparently very rare—although the describers commented that 'it is likely to be much more widespread' (Hasenpusch and Brock 2006). A female collected more recently has successfully founded a captive-breeding programme for *C. gargantua* at Museum Victoria.

However, the real richness of the insect fauna and its distribution is still somewhat speculative, but Platnick's (1991) contention that a 'pear-shaped' pattern of latitudinal distribution of the world's biodiversity, with greater representation in the southern than in the northern zones and richness diminishing from the tropics poleward is generally supported. It is very clear that levels of endemism across all insect orders in Australia are high and that the faunal elements present include (1) more ancient southern (Gondwanan, as above) lineages, largely restricted to the cooler southern areas and (2) more recent taxa as waves of immigration (and subsequent diversification) from the north, so that many of those taxa have predominantly Oriental affinities and are best represented in the north of Australia. These two major categories extend northward and southward, respectively. Especially where the Great Dividing Range creates elevational variety near the central east coast, the faunas can variously intermingle, so that latitudinal gradients in some taxa are reasonably clear—but also emphasising that the native insect faunas of different parts of Australia differ considerably and that many taxa are quite restricted in their geographical range. Further faunal divisions, and local endemism, occur from (1) the functional separation of the temperate southeast and southwest mainland areas by the Nullarbor Plain across central southern Australia, so that their faunas have evolved largely in isolation; (2) the separation of Tasmania from the mainland by Bass Strait, with relatively recent land bridges toward either end of the Strait; and (3) the vast arid and semi-arid 'desert' areas of central Australia, concentrating much of the fauna toward the more hospitable east coastal regions but enabling colonisation by some more tolerant taxa and largely isolating the insect faunas of the 'Top End' and north western coastal areas. Geographical and faunal differentiation is evident also from Papua New Guinea through the (Australian) Torres Strait Islands into the northern tip of Queensland, Cape York.

The above is rather simplistic, but helps to indicate the variety of faunal elements present. That variety, discussed by Austin et al. (2004) is thus a complex amalgam of southern and northern elements, leading them to comment that 'more difficult than estimating the size of the fauna is determining its origins and the factors involved in its evolution'. Whilst many of the elements can be recognised, many higher level (above species) groups commonly comprise mosaics of different elements that have interacted in complex ways. As examples, Austin et al. noted that most very species-rich insect families have components that can be assigned to all or most faunal elements. They cited Scarabaeidae, Tachinidae, Geometridae and Braconidae as representatives from different megadiverse orders each with a variety of faunal components. Some taxa do not fit clearly into any faunal group.

Recent influxes of alien insect species, introduced or otherwise adventive since European settlement constitutes another functional element superimposed on the native fauna.

Many native insect species are widespread, but many others are not—and some are very localised—but 'Australia exhibits all major elements of insect biodiversity' (Cranston 2009), with those elements almost invariably having high levels of endemism and richness. Loss of any local restricted habitats in terrestrial or inland water ecosystems may thus include losses of highly localised endemic resident insects.

6.3 Understanding Australia's Insect Richness

Approaches to estimating Australia's insect richness have involved combining numbers of described species and those otherwise known in collections, and informed estimates from specialists in each insect group. Total numbers are still imprecise, but continue to increase as discoveries increase and ideas are modified, from Taylor's polled total of 108,000 species (Taylor 1979, 1983), to the 140,000 of Nielsen and West (1991) and the almost 205,000 species estimated by Yeates et al. (2003). Uncertainties prevail. The assumptions and methodology of extrapolating from collected material to an entire fauna are sometimes unclear, different specialists can differ in their concepts of 'a species', which may differ also across taxa from wider consensus, and so on. Yeates et al. considered only 'morphological species', those that could be distinguished by some individual structural features and characteristics and so amenable to consistent categorisation by 'traditional' taxonomic methods. They thus did not include what they admitted may be substantial numbers of 'genetic species' whose inclusive effects on total numbers were unknown, but from examples of DNA barcoding elsewhere (p. 76) are likely to be large.

Taylor's (1983) approach was based on recognition of several 'levels' of insect species, as steps from (1) 'species in nature', then essentially subdivided to be represented by (2) 'species in collections' with these subsequently but gradually converted to (3) 'species fully named' and later included in broader revisions as (4) 'species taxonomically understood'. He sought advice from specialists on the presence of 'their' group(s) in Australia, in terms of percentages of species fully named, known but not yet named, and species in nature not yet collected. Comparable information on the insect faunas of North America and Britain (Fig. 6.1) revealed the substantial work needed for knowledge of the Australian fauna to approach data on these northern faunas, and demonstrated the very different information base for study and conservation.

Taylor also examined the progression through the above categories, recognising that survey and access rates (step 1–2, above), the 'naming rate' (step 2–3) and rate of taxonomic synthesis (step 3–4) all affect the transition of species in nature to species taxonomically understood through the broad 'rate of taxonomic activity'. He also noted, presciently, that knowledge gained with increased taxonomic understanding 'may cause us to revise radically our understanding of the numbers of species actually represented in collections or in nature' and that any additional biological information may cause taxonomists to revise their estimates of species numbers.

It remains to be clarified whether the numbers given by Yeates et al. (2003) are a realistic evaluation of Australia's insect richness—or whether even more recognisable species occur. In concluding their essay, those authors noted several then recent surveys of terrestrial insects that inferred higher numbers of species and that 'it is a significant challenge to directly compare results of those intensive surveys with estimates generated for broad taxonomic groups for the whole continent'. Needs to facilitate this included examination of host specificity among insect

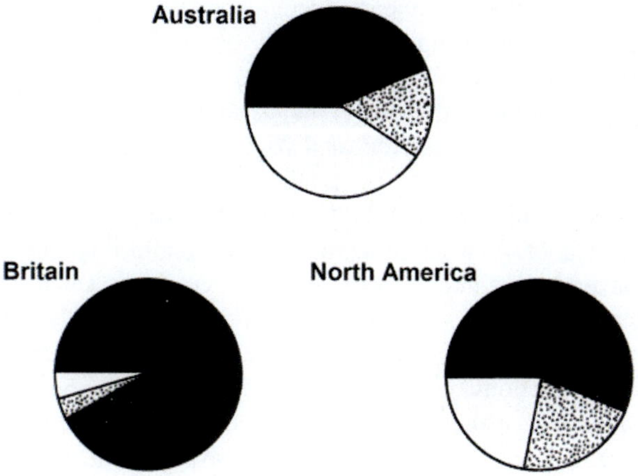

Fig. 6.1 An earlier assessment of the relative knowledge levels of the insect faunas of Britain, North America and Australia: the estimated proportion of species that are named (black), known but unnamed (dotted) and estimated yet to be collected (open). Information based on estimated totals of 23,500 species (Britain), 133,000 species (North America) and 108,000 species (Australia, known now to be a substantial underestimate, see text) (extracted from Taylor 1983, original ©. CSIRO)

herbivores in dominant plant communities [with this recognised as a key component of insect richness since Erwin's pioneering study (Erwin 1982) on neotropical forest beetles] and evaluating levels of beta- and gamma-diversity in different species guilds and ecosystems. Validation of any such estimates from any independent method is highly desirable, but also difficult.

For several large insect orders, Yeates et al. also gave estimates using the 'revision multiplier effect' (RM), essentially extrapolating from recent taxonomic revisions to help estimate the rate of species discovery and for comparison with estimates made by experts (E). For some taxa the two estimates were quite similar (Diptera: E 30000, RM 30744; Lepidoptera: E 20000, RM 20748), but larger differences also occurred, either upward or downward (Coleoptera: E 90000, RM 102596; Hymenoptera: E 43800, RM 28766). The biases of the RM approach are considerable, reflecting the numbers of revisions and of previous workers, but does provide some picture of current taxonomic research—and indicates that the rate of increase is not yet beginning to decrease to any indication that knowledge is nearing completion. Prospects for this to occur seem remote. Assuming that their estimates for numbers of terrestrial arthropod species were approximately correct, Yeates et al. noted that the historical record was of about 60,000 species described over some 230 years. At such a rate of description, the last species would not be described until the year 2700!

Over that long period many of those species will become extinct. Soberingly, many Australian insects in collections, both described and undescribed, are still known only from their type specimens or series of first-discovered individuals. Such scenarios span the entire period of taxonomic descriptions of Australian insects, from taxa collected on early expeditions to more recent discoveries, with corresponding increased label detail over that period, and also span all major regions of the continent. For some species, repeated targeted searches in and near the original discovery localities by specialists have failed to find them, and cases in which massive changes to the original collecting sites have occurred (many now unrecognisable in any form resembling their earlier condition) imply that the insects may indeed have disappeared. Nevertheless, proving that an insect species has become extinct in Australia is difficult, but can be claimed with reasonable confidence for the very few taxa currently regarded as such (p. 17). Several other formerly widely believed extinctions (based on the species not having been seen for many years and awareness that their former habitats have changed considerably in the intervening period) have been refuted by rediscoveries, either of small remnant populations within the historical range, of viable populations in other places, or of tantalising new records. The five following cases exemplify this. First, the Lord Howe Island stick insect (*Dryococelus australis*) was believed to have been exterminated from its isolated island habitat after the arrival of rats, and repeated searches failed to rediscover it until a few individuals were found on the isolated steep and difficult-to-access Balls Pyramid (Priddel et al. 2003). Progeny from a female taken to the Melbourne Zoo and raised through an innovative and initially tenuous captive rearing programme (Honan 2008) has enabled large numbers to be distributed more widely, and the stick insect is poised for reintroduction to Lord Howe after conclusion of a rat eradication campaign (Wilkinson 2014). Because stick insects from Lord Howe and Ball's Pyramid differ somewhat in appearance, doubts over whether the populations were conspecific were resolved through genetic analyses that confirmed this (Mikheyev et al. 2017). Second, the Ancient greenling damselfly (or Hemiphlebia damselfly, *Hemiphlebia mirabilis*), for long regarded as a 'living fossil' within the Odonata and of global interest was regarded as extinct because its few known small aquatic habitats in Victoria's Yarra Valley had succumbed to drainage and agricultural conversion during the middle of the twentieth century. Discovery of a population on a seasonal swamp on Wilsons Promontory, southern Victoria (Davies 1985) prompted renewed searches and study of this unusual species (Sant and New 1988), and it is now known to be more widely distributed in Victoria and adjacent south eastern South Australia, and in north-eastern Tasmania. Indeed, one population has recently been estimated to number at least a million individuals (Cordero-Rivera 2015). Third, Australia's unique 'Dinosaur ant' (*Nothomyrmecia macrops*) is known from only a small region of inland South Australia, following a somewhat fortuitous discovery by Taylor in 1977, and led to considerable international interest (Taylor 1978). Subsequent surveys revealed a number of colonies of *Nothomyrmecia* over a span of around 400 km (Watts et al. 1998). The ant had for long been known only from the original two specimens collected in Western Australia in the 1930s and where repeated searches over ensuing decades had failed to re-find it; despite this it was listed for protection in Western Australia in 1978. Fourth, the very primitive 'earwigfly', *Austromerope poultoni* (Mecoptera:

Meropeidae), is endemic to southern Western Australia. The single specimen collected in 1914 was not described until 1933, and the species was not seen again until 1962. Subsequent searches over 2001–2006 enhanced knowledge considerably, with 27 specimens collected across several sites, leading Abbott et al. (2007) to suggest that it might occur widely in the south-west forests and woodlands. The history mirrors that of many Australian insects: as Abbott et al. put it 'Although first collected accidentally nearly 100 years ago, the subsequent collection history of *A. poultoni* is one of serendipity'. Finally, the unusual endemic bee *Pharohylaeus lactiferus* (a Cloaked bee, Colletidae) had not been seen since 1923 until extensive searches in and around rainforest patches led Dorey (2021) to discover three isolated populations in Queensland. *P. lactiferus* apparently has quite specific foraging habits, and may nest in preformed cavities, as do many other hylaeine bees. The forest areas where it was found are threatened by continuing clearing and increasing risk of stochastic events such as fire, but the conservation status of the bee is still speculative.

The above species are each quite characteristic in appearance, and recognisable unambiguously. Many others are not as distinctive and are far more difficult to identify. 'To know and recognise each requires that we know something of what makes each species unique in comparison with all others' (Wheeler 2018), an undertaking that requires both descriptive taxonomy and 'scientific natural history' that can together help to create the understanding needed to describe insect diversity. Definitions of species (p. 72) vary, but few entomologists would query the reality that massive effort is needed to accelerate characterisation of Australia's insect faunal richness, should more definitive species numbers be needed for conservation advocacy. The largely under-documented diversity of some insect groups is illustrated well by the richness of ants in the arid regions of Australia, where Andersen (2016) hypothesised that movements of dune sands during warm glacial periods of the Pleistocene created an environment of highly fragmented biota confined to numerous isolated refuges—forming what he termed 'the perfect storm for speciation' among ants. Australia reportedly remained relatively warm during the glaciations that caused mass extinctions in the northern hemisphere and during periods of highest aridity up to about 85% of the continent was covered with sand dunes. Several ant genera have become 'megadiverse', with their richness only now starting to become apparent—and likely to reflect speciation only over the last few million years, so largely at the 'species-level' rather than through evolution of more strongly differentiated genera. Andersen (2016) showed that several ant genera each have more than 300 species in the arid zone, with the endemic genus *Melophorus* likely to contain more than 1000 species. Many ant genera there contain more than 100 species, and Andersen estimated that the total arid zone ant fauna in Australia could comprise up to 7500 species, rendering the Australian ant fauna among the world's richest—rivalling that of the Amazon basin, conventionally accepted as the richest.

Some workers have suggested that Australia's insect fauna is 'well-known' (Gaston 1991; Gaston and Hudson 1994), and Cranston (2009) commented that the taxonomic impediment has not deterred entomologists and conservation biologists from making substantial contributions to an understanding of regional insect biodiversity and its conservation. There is clearly still much to do, but it is also clear that changes to Australian environments over the last two centuries have adversely affected many insect habitats and the species they host. It is also clear that, despite lack of full documentation, levels of endemism among Australia's insects are high: Raven and Yeates (2007) opined that 'it is likely that about 90% of ... native arthropods ... are endemic'.

6.4 Endemism

Interpreting patterns of endemism amongst Australia's insects encompasses considering a wide range of historical and ecological factors and influences. Defining 'short-range endemism' as occupying a range of less than 10,000 km^2, Harvey (2002) reported that several non-insect invertebrate groups consisted primarily of such species. Many of these were linked by having poor dispersal powers and being restricted to discontinuous habitats; highlighted taxa were examples from Gastropoda, Oligochaeta, Araneae (mygalomorph spiders), Schizomida, Diplopoda, Phreatoicidea and Decapoda (freshwater crayfish). In contrast to many claims of narrow range restrictions, some based on inadequate and infrequent surveys or simple taxonomic misinterpretations, several of the above groups have been investigated sufficiently to confirm the reality of those distributions, and the numbers of species involved in such taxa continue to increase. The Onychophora, velvetworms (or '*Peripatus*', many of which are Gondwanan relicts in Australia), includes some species restricted to single localities in which permanently moist conditions prevail and where their poor mobility has allowed genetic differentiation to furnish distinctive and extremely localised evolutionary units. Relatively few higher taxonomic groups show this extreme level of endemism although their individual species can do so. For example, a predominant family of water beetles (Dytiscidae) is widespread, but a number of species restricted to particular cave systems as specialised stygofauna are among the narrowest of all narrow range endemics. In assessing the assemblages of these water beetles, each characteristic of a single calcrete aquifer in Western Australia, Cooper et al. (2002) likened these aquifers to 'subterranean islands with independently evolved taxa'; such restricted biota can readily become threatened if their environments are disturbed. Refugial habitats, such as elevated mountain areas and sites that have withstood widespread aridification and forest contraction, caves and subterranean systems, as well as individual river systems or drainage basins, provide arenas in which such species can persist. Overview of the Odonata (Hawking and Theischinger 2004) helped to indicate the likely levels of endemism among insect groups: of the 110 genera then recorded from Australia, 49 are endemic, and 22 also monotypic. Sixteen of the latter were included on a list of 27 species regarded as 'Near Threatened', mostly through having very restricted ranges and being confined to specialised habitats such as streams or bogs. As with many other insect groups—most having received far less attention than Odonata—the high proportion (ca 30%) of Data Deficient species was regarded as 'possibly the major concern facing the Australian Odonata conservation' (Hawking and Theischinger 2004), as thwarting a balanced overview of needs beyond those of individual species. The category 'Data Deficient', indeed, is necessarily predominant for insects of most orders in Australia.

Assessment of endemism clearly relates to the scale of consideration and the extent of investigation invested in confirming (or otherwise) that condition. Insects simply endemic to Australia may be very widely distributed across the country, or much more restricted, with the latter the greater concern in conservation. A narrow-range endemic species may be restricted in space, defined by geographical or elevational coordinates, or by ecological conditions as an equivalent determinant of availability of a critical resource such as a particular food plant. For either, areas of endemism are irreplaceable. Those with higher numbers of endemic taxa or that harbour especially unusual lineages of insects deserve the highest priority for protection and conservation.

The following two examples illustrate these scenarios:

1. For flightless rainforest insects in Australia's Wet Tropics (Fig. 6.2), Yeates et al. (2002) distinguished two nested levels as species that are 'regional endemics' (confined to the Wet Tropics) and 'subregional endemics' (confined to one of the 14 upland rainforest regions surveyed), interpreting data accumulated across the region over some 20 years and from about 350 sites. The total number of regional endemic species (274) included members of 15 families, with highest richness in Hemiptera: Aradidae (46 species) and three families of Coleoptera (Carabidae 78, Scarabaeidae 32, Tenebrionidae 41) together included 137 subregional endemics. The latter numbers differed considerably (from 0 to 28) across different subregions, with the greatest numbers in those subregions with highest overall species richness. For the 14 subregions, the percentage of subregional endemics ranged over 0–81%.

2. Radiations of native flora are associated with parallel radiations of some groups of plant-feeding insects, many of which are host-specific and thereby ecological specialists. Several genera of Psylloidea (Hemiptera), for example, have diversified as independent parallel radiations on the two most diverse tree genera, namely *Eucalyptus* s.l. (Myrtaceae, with about 700 species) and *Acacia* s.l. (Mimosaceae, with >950 species), and some other taxa among phytophagous Hemiptera, Coleoptera and Lepidoptera are still to be explored more comprehensively (New 1983).

In addition to host plant specificity, many of these insect herbivores are not coextensive with their hosts, but occur over only small parts of the host plant range. As Yen (2002) noted 'short-range endemics' among such herbivores can be distinguished on either (1) strict host plant specificity with a narrowly distributed plant, or (2) narrow distribution on a host plant with a broad geographical distribution. The former may be especially prone to co-extinction should that host disappear. Examining the radiations of the psylloid genera *Glycaspis* (with 128 of the 132 species then described found on *Eucalyptus*) and *Acizzia* (77 morphospeces, mostly

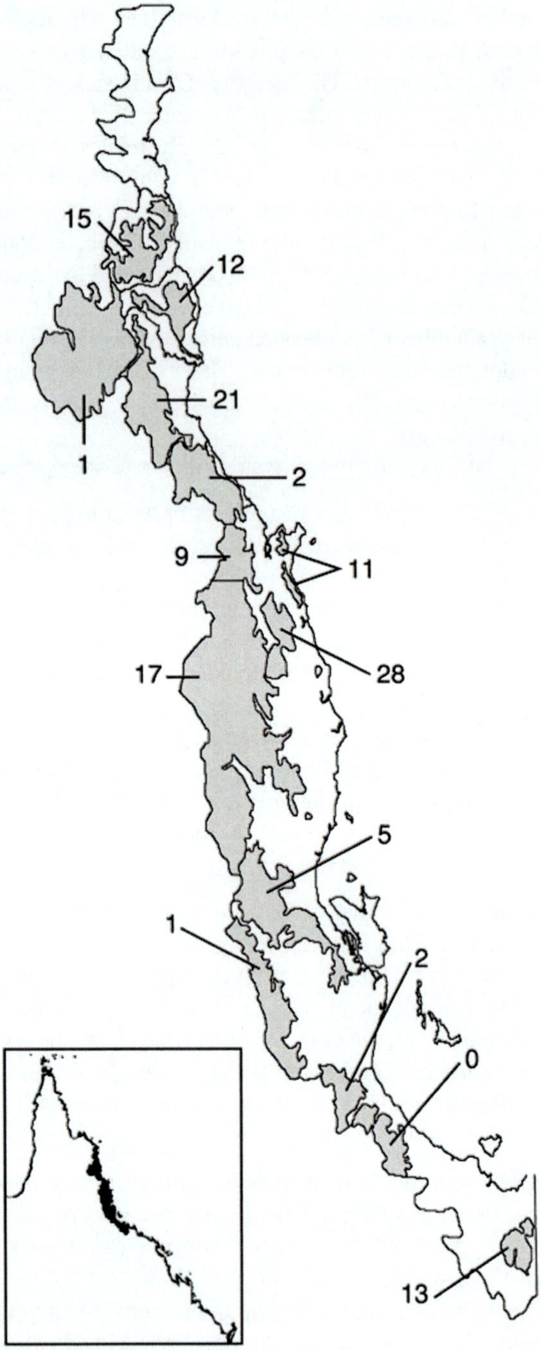

Fig. 6.2 The Australian Wet Tropics region of Queensland. The 14 upland subregions (>300 m) are shown, and for each the number of flightless endemic insect species is given (after Yeates et al. 2002)

undescribed, on *Acacia*), Yen found most species of both had been reported from a single host (*Glycaspis* 69 species, *Acizzia* 41) with only low numbers of species each from more than five hosts (*Glycaspis* 6, *Acizzia* 11)—so that apparent host specificity was high. Increased intricacy of the association tended to increase specificity, so that the gall-inducing species of *Glycaspis* were all restricted to a single host. Conversely, some free-living species used several different hosts in places across a broad range. Additional descriptions of new species of *Acizzia* from host plants that are themselves listed as threatened because of their narrow ranges have further implied their vulnerability to coextinctions (Taylor and Moir 2009). Thus, the host of *Acizzia vedi* (*Acacia veronica*) has a very narrow distribution in Western Australia, where putative threats include clearing, inappropriate fire regimes and climate change. Six populations of the acacia are known, but the psyllid has been found on only one of these, so may be far more endangered than its host.

This second example demonstrates that conserving a plant host is a prerequisite for conserving a 'dependent species', in this case a monophagous herbivorous insect, but this may not be the only need for that insect (Moir et al. 2010, 2011, 2014). In many cases, knowledge of the host species or population is inadequate to optimise actions needed to prevent any dependent species from becoming extinct—indeed, as Moir et al. (2010) noted, some conservation actions for the host may actually be detrimental to a dependent insect. Translocation of plants to an area where the dependent insect does not occur is one such case. Most insect herbivores are not considered in such exercises, although butterflies provide exceptions in which appreciation of specific ecological associations has transformed the practice of scientifically informed insect conservation (Merckx et al. 2013). The intricate tripartite relationships between some lycaenid butterflies, specific larval food plants and particular mutualist ant species have attracted much attention, and confirmed that each of the 'partners' may need separate specialised attention to secure them, and to continue their interactions. Such interacting taxa constitute a 'community module' (sensu Mouquet et al. 2005), but the different taxa can sometimes be given different forms of individual priority. As for the Large blue butterflies (*Maculinea* spp.) in Europe (Hayes 2015), in this context the butterfly may be thought of as an 'umbrella species' and the mutualist ant as a 'keystone species'.

'Endemism' is thus a concept with many nuances of interpretation, but all these at some level imply that losses impoverish biota by the removal of unique and irreplaceable localised components—here, insect species. Anthropogenic changes to Australian environments are creating narrow-range endemics (Harvey 2002). In consequence, as these are noticed they may be noted among those species of conservation attention and gain closer study of their conservation status. As a wider case of functional significance of endemisms, local endemism was the basis for designating and ranking 'critical faunas' for the world's milkweed butterflies (Nymphalidae: Danainae) by Ackery and Vane-Wright (1984). The approach involves selection of the sites (or larger geographical regions) ranked by numbers of endemic species, commencing with the greatest number of endemic taxa, and complementing by successive additions until all species were represented on at least one site. That approach is suitable only for very well understood taxa, but can give clear indication of the 'most significant' places in which to concentrate conservation for the greatest 'savings' in terms of numbers of unique taxa present. Somewhat parallel in intent, but without strict complementarity, the selection of Prime Butterfly Areas in Europe (Van Swaay and Warren 2003) was based on meeting three broad criteria: European range-restricted species, threatened species (defined by the IUCN criteria) and species listed in the Bern Convention and/or the European Union Habitats and Species Directive. Such exercises in selecting priority sites for insect conservation on representation and/or diversity of focal groups clearly depend on sufficient knowledge to do so, including reliable distributional information.

However, for many plant-associated insect groups in Australia, presumption of narrow ranges is largely due to undersurvey and inadequate taxonomic penetration and, although likely, is not yet able to be confirmed from rigid investigation, pending which some doubts can easily be raised. The moth family Oecophoridae contains numerous species with larvae specialised to feed on dead foliage of eucalypts and related Myrtaceae in forests and woodlands, where they are regarded as significant contributors to the breakdown of this tannin-rich litter on the ground. The family has for long been recognised as very diverse—easily the most diverse family of Lepidoptera in Australia—with almost all species endemic and many known only from rearing from samples of leaf litter from single or few localities. Common (1994) estimated the total Australian oecophorid fauna to exceed 5000 species, and at that time only about 1850 valid species names were available. Local richness can be high, evidenced by numbers of species emerging from single samples of litter, but estimating and characterising local diversity cannot be achieved realistically, and the extent of further sampling and species-level taxonomic work to clarify the roles of this functionally important group remains daunting, and unlikely.

Similar comments apply to other species-rich insect herbivore groups in Australia. Reflecting the complexities of recent radiations and diversification, many such closely related species are very similar in appearance, and can probably be separated most consistently by DNA methods (p. 173); for most taxa, this analysis has not yet been undertaken. The impracticability of such subtle species recognition helps to emphasise need for a wider perspective for conservation.

Nevertheless, these groups also help to highlight the complexities of insect diversity that are often not heeded by non-entomologists, and also that the numbers of endemic species may actually by far exceed current estimates, and their richness contribute to the values of a site or biotope. The insects of all Australia's major natural habitat categories include numerous endemic taxa, as summarised for forests (New 2018), grasslands (New 2019) and inland waters (New 2021). In all these varied systems, much of that diversity can easily become threatened, with many external impacts usually not considered for their effects on insects.

Setting priorities in order to best promote conservation interest and progress amongst the diverse arrays of Australian insects that await description and clear recognition demands alternatives to simply working according to the whims and availability of relevant experts, however valuable those contributions continue to be. Surveys of insects and other invertebrates in Australia can rarely approach completeness. Selection of particular taxa for attention means that the primary focus can be on the taxon itself, or consider wider evolutionary or ecological context. For practicality of increasing insect inclusions in conservation-related assessments, New (1995, 2009a, b) suggested that insects fall into three main categories, as (1) 'well-known groups', for which taxonomic and other information is already sufficient for meaningful assessment, conservation and understanding: butterflies are the paramount example; (2) 'catch-up groups', for which strong basic understanding and documentation already exists, and for which limited but concentrated work could transform them to 'well-known'; and (3) 'black hole groups', for which information is fragmentary, the taxonomic impediment is greatest, and the chances of redressing this to render these groups more 'tangible' are remote—however useful it may be to demonstrate their vast richness and ecological complexity, that information is unlikely to stray beyond the ambit of the few specialists involved. Nevertheless, exceptions occur. Recent work on the endemic gall-forming flies (Fegusoninidae) on eucalypts has revealed unexpectedly high richness and complexity and facilitated their consideration in conservation (Purcell et al. 2016). As another example, many Sun-loving moths are the specific pollinators of individual species of Rutaceae, and their larvae feed on developing seeds of that obligatory host plant so that interactions beween moth and host are very intricate. Both Fergusoninidae and Heliozelidae are widely distributed families in Australia.

Harvey et al. (2011) suggested concentrating conservation interest on groups that are largely narrow-range endemics, on the basis of focusing efforts on the groups that are most at risk, most likely to be undescribed, and most likely to have distributions so small that development proposals could affect their entire range. They commented that if government agencies are to reduce the impacts of human activities, information on those species that are most at risk provides considerable value, with accurate identification of individual species and detailed documentation of their range and habitat requirements contributing much to that value. Decisions of this nature have been important in Western Australia in response to discoveries of unique and novel troglobitic and subterranean arthropod radiations and camaenid land snails, enabling their inclusion in mining and other exploration proposals.

In principle, any such discovery merits conservation attention. In practice, many do not command that interest and the recognition of novelty is itself highly uncertain, depending on whether the discovery is by (or brought to the attention of) a worker versed in the insect group involved or concerned for its environment or place of discovery. Such association, particularly for the less popular insect orders, is largely fortuitous.

References

Abbott I, Burbidge T, Wills A (2007) *Austromerope poultoni* (Insecta, Mecoptera) in south-West Western Australia: occurrence, modelled geographical distribution, and phenology. J R Soc W A 90:97–106

Ackery PR, Vane-Wright RI (1984) Milkweed butterflies: their cladistics and biology. British Museum (Natural History). Academic Press, London

Andersen AN (2016) Ant megadiversity and its origins in arid Australia. Austral Entomol 55:132–137

Austin AD, Yeates DK, Cassis G, Fletcher MJ, La Salle J et al (2004) Insects 'down under' – diversity, endemism and evolution of the Australian insect fauna: examples from select orders. Aust J Entomol 43:216–234

Ball JB (2012) Lepidopterology in southern Africa: past, present and future. In: New TR (ed) Insect conservation: past, present and prospects. Springer, Dordrecht, pp 279–300

Bossart JL, Carlton CE (2002) Insect conservation in America: status and perspectives. Am Entomol 48:82–92

Braby MF (2000) Butterflies of Australia. Their identification, biology and distribution (2 vols). CSIRO Publishing, Melbourne

Braby MF (2016) The complete field guide to butterflies of Australia, 2nd edn. CSIRO Publishing, Melbourne

Common IFB (1994) Oecophorine genera of Australia I. The *Wingia* group (Lepidoptera: Oecophoridae). Monographs on Australian Lepidoptera. CSIRO Publishing, Collingwood

Contador TA, Kennedy JH, Rozzi R (2012) The conservation status of southern South American aquatic insects in the literature. Biodivers Conserv 21:2095–2107

Cooper SJB, Hinze S, Leys R, Watts CHS, Humphreys WF (2002) Islands under the desert: molecular systematics and evolutionary origins of stygobitic water beetles (Coleoptera: Dytiscidae) from Central Western Australia. Invert Syst 16:589–598

Cordero-Rivera A (2015) Demographics and adult activity of *Hemiphlebia mirabilis*: a short-lived species with a huge population size (Odonata: Hemiphlebiidae). Insect Conserv Divers 9:108–117

Cranston PS (2009) Biodiversity of Australian insects. In: Foottit RC, Adler PH (eds) Insect biodiversity: science and society. Wiley-Blackwell, Chichester, pp 83–105

Davies DAL (1985) *Hemiphlebia mirabilis* Selys: some notes on distribution and conservation status (Zygoptera: Hemiphlebiidae). Odonatologica 14:331–339

Dorey JB (2021) Missing for almost 100 years: the rare and potentially threatened bee, *Pharohylaeus lactiferus* (Hymenoptera, Colletidae). J Hymenopt Res 81:165–180. https://doi.org/10.3897/jhr.81.59365

Erwin TL (1982) Tropical forests: their richness in Coleoptera and other arthropod species. Coleopt Bull 36:74–75

Gaston KJ (1991) The global magnitude of insect species richness. Conserv Biol 5:283–296

Gaston KJ, Hudson E (1994) Regional patterns of diversity and estimates of global insect species richness. Biodivers Conserv 3:493–500

Harvey MS (2002) Short-range endemism among the Australian fauna: some examples from non-marine environments. Invertebr Syst 16:555–570

Harvey MS, Rix MG, Framenau VW, Hamilton ZR, Johnson MS et al (2011) Protecting the innocent: studying short-range endemic taxa enhances conservation outcomes. Invertebr Syst 25:1–10

Hasenpusch J, Brock PD (2006) Studies on the Australian stick insect genus *Ctenomorpha* Gray (Phasmida: Phasmatidae: Phasmatinae), with the description of a new large species. Zootaxa 1282:1–15

Hayes MP (2015) The biology and ecology of the large blue butterfly *Phengaris* (*Maculinea*) *arion*: a review. J Insect Conserv 19:1037–1051

Hawking JH, Theischinger G (2004) Critical species of Odonata in Australia. Int J Odonat 7:113–132

Henning S, Henning GA (1989) South African red data book: butterflies. Foundation for Research Development, Council for Scientific and Industrial Research, Pretoria

Honan P (2008) Notes on the biology, captive management and conservation status of the Lord Howe Island stick insect (*Dryococelus australis*) (Phasmatodea). J Insect Conserv 12:399–413

Kristensen NP, Hilton DJ, Kallies A, Milla L, Rota J et al (2015) A new extant family of primitive moths from Kangaroo Island, Australia, and its significance for understanding early Lepidoptera evolution. Syst Entomol 40:5–16

Marks EN (1991) Biographical history. In: Naumann ID (ed) The insects of Australia. Melbourne University Press, Melbourne, pp 198–220

McGuiness CA (2001) The conservation requirements of New Zealand's nationally threatened invertebrates. Threatened species occasional publication no 20. Department of Conservation, Wellington

Mecenero S, Ball JB, Edge DA, Hamer ML, Henning GA et al (2013) Conservation assessment of butterflies of South Africa, Lesotho and Swaziland: red list and atlas. Saftronics, Johannesberg and Animal Demography Unit, Cape Town

Mecenero S, Edge DA, Staudte HS, Coetzer BH, Coetzer AJ et al (2021) Outcomes of the southern African Lepidoptera conservation assessment (SALCA). Metamorphosis 31:1–160

Merckx T, Huertas B, Basset Y, Thomas J (2013) A global perspective on conserving butterflies and moths and their habitats. In: Macdonald DW, Willis KJ (eds) Key topics in conservation biology. Wiley, Chichester, pp 239–257

Mikheyev AS, Zwick A, Macgrath MJL, Grau ML, Qiu L et al (2017) Museum genomics confirms that the Lord Howe Island stick insect survived extinction. Curr Biol 27:3157–3161

Moir ML, Vesk PA, Brennan KE, Keith DA, Hughes L, McCarthy MA (2010) Current constraints and future directions in estimating coextinctions. Conserv Biol 24:682–690

Moir ML, Vesk PA, Brennan KEC, Keith DA, McCarthy MA, Hughes L (2011) Identifying and managing threatened invertebrates through assessment of coextinction risk. Conserv Biol 25: 787–796

Moir ML, Hughes L, Vesk PA, Leng MC (2014) Which host-dependent insects are most prone to coextinction under changed climates? Ecol Evol 4:1295–1312

Moore NW (1997) Dragonflies: status survey and conservation action plan. IUCN/SSC Odonata Specialist Group, IUCN, Cambridge

Mouquet N, Belrose V, Thomas JA, Elmes GW, Clarke ROP, Hochberg ME (2005) Conserving community modules: a case study of the endangered lycaenid butterfly *Maculinea arion*. Ecology 86:3160–3173

Naumann ID (ed) (1991) The insects of Australia. Melbourne University Press, Carlton

New TR (1983) Systematics and ecology: reflections from the interface. In: Highley E, Taylor RW (eds) Australian systematic entomology: a bicentenary perspective. CSIRO Publishing, Melbourne, pp 50–79

New TR (1995) Introduction to invertebrate conservation biology. Oxford University Press, Oxford

New TR (2009a) Insect species conservation. Cambridge University Press, Cambridge

New TR (2009b) Habitat offsets for insect species conservation: practicality or placebo? J Insect Conserv 13:139–141

New TR (2011) 'In considerable variety': introducing the diversity of Australia's insects. Springer, Dordrecht

New TR (2018) Forests and insect conservation in Australia. Springer, Cham

New TR (2019) Insect conservation and Australia's grasslands. Springer, Cham

New TR (2021) Insect conservation and Australia's inland waters. Springer, Cham

New TR, Samways MJ (2014) Insect conservation in the southern temperate zones: an overview. Austral Entomol 53:26–31

New TR, Yen AL (2012) Insect conservation in Australia. In: New TR (ed) Insect conservation: past, present and prospects. Springer, Dordrecht, pp 193–212

Nielsen ES, West JC (1991) Biodiversity research and biological collections: transfer of information. In: Forey PL, Humphrey CJ, Vane Wright RI (eds) Systematics and conservation evaluation. Clarendon Press, Oxford, pp 101–121

Patrick B, Patrick H (2012) Butterflies of the South Pacific. Otago University Press, Dunedin

Platnick NI (1991) Patterns of biodiversity: tropical vs temperate. J Nat Hist 25:1083–1088

Priddel D, Carlile N, Humphrey M, Fellenberg S, Hiscox D (2003) Rediscovery of the 'extinct' Lord Howe Island stick insect (*Dryococelus australis* (Montrouzier) (Phasmatodea)) and recommendations for its conservation. Biodivers Conserv 12:1391–1403

Purcell MF, Wallenius TC, Yeates DK, Rowell DM (2016) Larval dorsal shield morphology is highly correlated with gall type in the enigmatic gall-forming fly, *Fergusonina* Malloch (Diptera: Fergusoninidae). Aust J Zool 64:233–248

Pyle RM (2012) The origins and history of insect conservation in the United States. In: New TR (ed) Insect conservation: past, present and prospects. Springer, Dordrecht, pp 157–170

Raven PH, Yeates DK (2007) Australian biodiversity: threats for the present, opportunities for the future. Aust J Entomol 46:177–187

Rentz DCF (1980) A new family of ensiferous Orthoptera from the coastal sands of Southeast Queensland, Australia. Mem Qld Mus 20:49–64

Rentz DCF (1994) Orthopteroid insects in threatened habitats in Australia. In: Gaston KJ, New TR, Samways MJ (eds) Perspectives on insect conservation. Intercept, Andover, pp 125–138

Samways MJ (2008) Dragonflies and damselflies of southern Africa. Pensoft, Sofia

Samways MJ, Hamer M, Veldtman R (2012) Development and future of insect conservation in South Africa. In: New TR (ed) Insect conservation: past, present and prospects. Springer, Dordrecht, pp 245–278

Sands DPA, New TR (2002) The action plan for Australian butterflies. Environment Australia, Canberra

Sant GJ, New TR (1988) The biology and conservation of *Hemiphlebia mirabilis* Selys (Odonata: Hemiphlebiidae) in southern Victoria, Technical report series, no 82. Arthur Rylah Institute for Environmental Research, Melbourne, pp 1–35

Stewart AJA, New TR (2007) Insect conservation in temperate biomes: issues, progress and prospects. In: Stewart AJA, New TR, Lewis OT (eds) Insect conservation biology. CAB International, Wallingford, pp 1–33

Taylor RW (1978) *Nothomyrmecia macrops*: a living fossil ant rediscovered. Science 201:979–985

Taylor RW (1979) Some statistics relevant to Australian insect taxonomy. CSIRO Div Entomol Rep 8:1–9

Taylor RW (1983) Descriptive taxonomy: past, present, and future. In: Highley E, Taylor RW (eds) Australian systematic entomology: a bicentenary perspective. CSIRO Publishing, Melbourne, pp 93–134

Taylor GS, Moir ML (2009) In threat of co-extinction: two new species of *Acizzia* Heslop-Harrison (Hemiptera: Psyllidae) from vulnerable species of *Acacia* and *Pultenea*. Zootaxa 2249:20–32

Theischinger G, Hawking J (2021) The complete field guide to dragonflies of Australia, 2nd edn. CSIRO Publishing, Clayton South

Upton MS (1997) A rich and diverse fauna. The history of the Australian National Insect Collection, 1926-1991. CSIRO Publishing, Melbourne

Van Swaay C, Warren M (eds) (2003) Prime butterfly areas in Europe: priority sites for conservation. National Reference Centre for Agriculture, Nature and Fisheries, Ministry of Agriculture, Nature Management and Fisheries, the Netherlands

Watts CHS, McArthur A, Foster R (1998) Notes on the distribution of the dinosaur ant *Nothomyrmecia macrops* Clark (Hymenoptera: Formicidae) in South Australia. Aust Entomol 25:29–31

Watts C, Stringer I, Gibbs G (2012) Insect conservation in New Zealand: an historical perspective. In: New TR (ed) Insect conservation: past, present and prospects. Springer, Dordrecht, pp 213–243

Wheeler Q (2018) Blank canvas: the case for descriptive taxonomy. Integr Comp Biol 58:1118–1121

Wilkinson R (2014) Return of the phasmid. Media Dynamics, Hamilton Central

Yeates DK, Bouchard P, Monteith GB (2002) Patterns and levels of endemism in the Australian wet tropics rainforest: evidence from flightless insects. Invertebr Syst 16:605–619

Yeates DK, Harvey MS, Austin AD (2003) New estimates for terrestrial arthropod species-richness in Australia. Rec South Aust Mus Memoir series 7:231–241

Yen AL (2002) Short-range endemism and Australian Psylloidea (Insecta: Hemiptera) in the genera *Glycaspis* and *Acizzia* (Psyllidae). Invertebr Syst 16:631–636

Chapter 7
Levels of Concern and Approach

7.1 Introduction

As elsewhere, insect conservation in Australia has developed through two main themes. The more familiar is focus on individual species perceived as threatened and in danger of extinction, and which have been selected in some way for appraisal of conservation needs and management. The second, complementary, approach is direct attention to diversity, most frequently through insect richness in particular habitats or biotopes, or in selected sites that become threatened by human activities. The approaches can overlap, as when individual species become flagships for restricted vegetation types or wetlands that support far wider characteristic biodiversity. Those species then become ambassadors for wider conservation interests. Biogeographical considerations may also become important in detecting 'centres' of richness, perhaps in conjunction with vegetation; thus, the forests of northern and eastern Australia are recognised as harbours for high insect diversity and endemism. For butterflies, the isopleths of species numbers devised by Kitching and Dunn (1999) revealed the highest richness in those regions and also the gradient trends toward lower richness elsewhere. The major east-west decline reflected the western edge of the Great Dividing Range separating the moister environments of the east coast from the arid interior of the continent, with consequent attenuation of vegetational and butterfly variety. Elevational and latitudinal gradients also influence butterfly richness. Such regional differences occur also in other insect groups, such as ants, for which genus richness tends to be highest in vegetationally rich forest regions in the tropics and near tropics, but with some of the fewer genera in the arid region becoming very diverse.

These twin levels of concern, single focal species (commonly referred to as 'the fine-filter approach') and the wider community or biotope focus (the 'coarse-level' or 'coarse-filter approach': Chap. 10) are complementary, but with interest in the latter inevitably increasing with the realisation of the impracticalities of relying on the former for any comprehensive success. However, almost universally, focus on

© The Author(s), under exclusive license to Springer Nature Switzerland AG 2022

T. R. New, *Insect Diversity, Declines and Conservation in Australia*, Fascinating Life Sciences, https://doi.org/10.1007/978-3-030-90134-9_7

single 'species' (or subspecies) provides a tangible target for conservation attention, as an enterprise that can be readily understood by managers and communicated easily to a wider lay constituency. In contrast, conservation of 'communities' or other coarse-level entity can appear far more nebulous and complex. Much insect conservation is indeed 'species-led'. It thus becomes important to ensure that every such insect ambassador for conservation is selected responsibly, and that conservation measures are seen to be credible, understandable and sensible.

7.2 Listing for Priority

Attention to both single species and wider insect assemblages and communities in Australia is accompanied by formal notice of threatening processes, so that a number of key actual and potential threats to the Australian environment have some potential for actions to prevent or alleviate their impacts. As in many other places, Australia's insects may attract formal notice of conservation need by becoming individually 'listed' to in some way reflect that they are 'threatened' and consequently be signalled for conservation need. Listing may occur nationally under the Commonwealth's Environment Protection and Biodiversity Conservation Act (the EPBC Act) and/or under the individual legislations of range States and Territories. Reflecting the great variety of environments across the vast continent and patterns of endemism, most species of individual concern are restricted to a single State or Territory or a single outlying territory. More rarely, such a species occurs in two or more adjacent areas: thus, the Golden sun-moth (*Synemon plana*, Castniidae) on the south-east mainland is listed under EPBC and the three separate legislations of Victoria, New South Wales and the Australian Capital Territory. In such cases, non-uniformity over obligations from listing can lead to confusions or inconsistencies in response and reflect differences in conservation status across a species' range. However, only a tiny proportion of deserving species have been noticed in this way—a recent overview (Taylor et al. 2018) recorded only 28 invertebrates on Australian threatened species lists and/or the global IUCN Red List.

Whereas legislative 'listing' a species may confer obligation or priority for conservation investigation or action, the converse is also a serious consequence—especially in view of the very small numbers of listed taxa. Namely, 'species that are not listed are not protected and no legal mechanism exists to prevent destruction or degradation of their habitats' (Walsh et al. 2013). Although it is assumed widely that there is some 'automatic' link between listing a species and attempting to conserve it, this is far from so (Farrier et al. 2007), and outcomes can vary greatly across different legislations and interests. However, where formal obligation arises, a decision to list a species is implicitly also a decision to manage that species. In most situations, supporting resources for this are very limited, and in practice impose some form of triage in selecting which species are attended. The implied nexus between listing and conservation thereby does not guarantee that a listed species will take practical priority over others or even receive any attention. The three-staged perspective offered by Farrier et al. reflects this broader view of the relative values involved as illustrated through New South Wales legislation. The first stage (recovery planning) examines the conservation needs of a species across its range, the threats present or anticipated, and the actions needed to stop further deterioration in conservation status and to promote recovery—but without guarantee that any such actions will be undertaken. The second stage (addressing issues of resource allocation between all listed species) involves partitioning the demands on resources across all the relevant recovery plans, examining overlapping needs and opportunities for implementation, again without commitment to pursuing these. Third, threatened species conservation is integrated into the strategic land use planning process, determining the extent to which the wider community is committed to conserving threatened species habitat over competing priorities, and who will bear the cost.

The great majority of animals listed in Australia are 'charismatic' vertebrates for which public concern is widespread and can be garnered easily. The situation can differ markedly for invertebrates that, without such popular support, are likely to escape attention. As Hutchings (2004) succinctly summarised it, the dearth of invertebrate nominations reflects that 'they are small and largely hidden from view and the public's perception of invertebrates as having little value'. Further nominations can be made by any concerned person or group. These largely bias toward more popular taxa (with butterflies dominating most such listings in Australia), and so away from less popular taxa, and the relative representation of invertebrates decreases further. In short, lists of protected species are taxonomically unbalanced and do not adequately reflect the needs of Australian fauna (Farrier et al. 2007). However, any reasonably representative listing of threatened insects in Australia would rapidly become unwieldy through the large numbers involved, and—whilst conveying realistic information on the scope of need—would simply 'swamp' any chances of broad conservation progress. That so many species are needy simply corroborates the urgency and scale of those needs. Chances of redressing this fully are remote, but drawing formal attention to the numbers of threatened insect species remains a key conservation need, and the scale of need emphasises both the inadequacy of resources and attention given and the urgency of redress.

Rather than continue with the currently predominant *ad hoc* approach of opportunistically listing insect species as they are nominated responsibly and their status and needs reviewed carefully, a more considered and strategic selection of taxa for listing deserves serious consideration (Taylor et al. 2018), to focus more on community support, ecological representation of the listed taxa and acknowledgement of taxonomic diversity through a credible, responsible and integrated selection.

Indeed, Hutchings (2004) advocated that listing of invertebrates in Australia is suitable only for the most iconic species that can become flagship taxa as 'the invertebrate equivalent to the koala', in order to raise community awareness. Listing others was regarded as inefficient. Otherwise, efforts are more constructively directed to ensuring protection of all major habitats and—by default—their associated invertebrate fauna, both known and unknown.

'Iconic species' are those that have important social and/or scientific values. Taylor et al. (2018) noted that those values could include the social attributes of aesthetic value, appeal to the media, cultural significance, and having values for tourism or economic impact. Scientific values also vary, but could include key relictual status, being phylogenetically isolated, having particular ecological importance, or others. In addition, 'threatened species' comprise a second category of flagship taxa—and a species may have both social and scientific importance—but have the major roles of value in increasing awareness of insect importance, vulnerability and conservation needs. Taylor et al. also urged that, very broadly, such flagship species should be selected 'for visual appeal, ecological importance, scientific and/or social value and, above all, resonance with the general public', and should be promoted as representing much wider biodiversity and so contribute directly to the conservation management of habitat and ecological communities.

7.3 Extending Ecological Coverage

7.3.1 Bioregions

The ecological variety of Australia's environments, each with characteristic suites of insects, so many of them localised endemics, is reflected in designation of the nation's 89 geographical 'bioregions' and the representation of the few listed species of insects and their allies (Fig. 7.1). This system was anticipated to provide a realistic overview of ecological variety and was based initially on comprehensive analysis of Australia's ecological communities as the Interim Biogeographic Regionalisation for Australia (IBRA) (Thackway and Cresswell 1995). This partitioning arose from earlier systems of subdividing Australia that commenced with the tripartite division into 'Torresian', 'Eyrean' and 'Bassian' regions proposed by Spencer (1896) (as in Fig. 7.2 with the later addition of the south west mainland as a distinct Bassian region), and which for long guided biogeographical interpretations of the Australian biota, with continuing debate over the most suitable ways to partition the large arid/semiarid Eyrean (Eremaean) region (Ebach and Murphy 2020). The modern bioregions collectively include 419 'subregions' (Australian Government 2017) and the system has been endorsed as a key national policy framework to link with the National Reserve System (NRS) to define priorities for the long-term protection of Australia's biodiversity and assure that all key systems are represented. The NRS arose from the ideal of attaining a 'Comprehensive, Adequate and Representative' (CAR) reserve system (Australian Government 2017). The government target is to protect 17% of the terrestrial environment under the NRS.

A somewhat more simplistic categorisation of major habitat types based on predominant vegetation features indicates these broader ecological units (Fig. 7.3).

A key focus of a proposed strategy to hone effectiveness of insect conservation in Australia is to use the bioregion system as a guide to selecting and nominating flagship insect species for formal listing (Taylor et al. 2018) to increase geographical and ecological coverage of the ecosystems present. A major geographical bias was evident from the existing listed species: 36 of the bioregions, mostly in the arid or

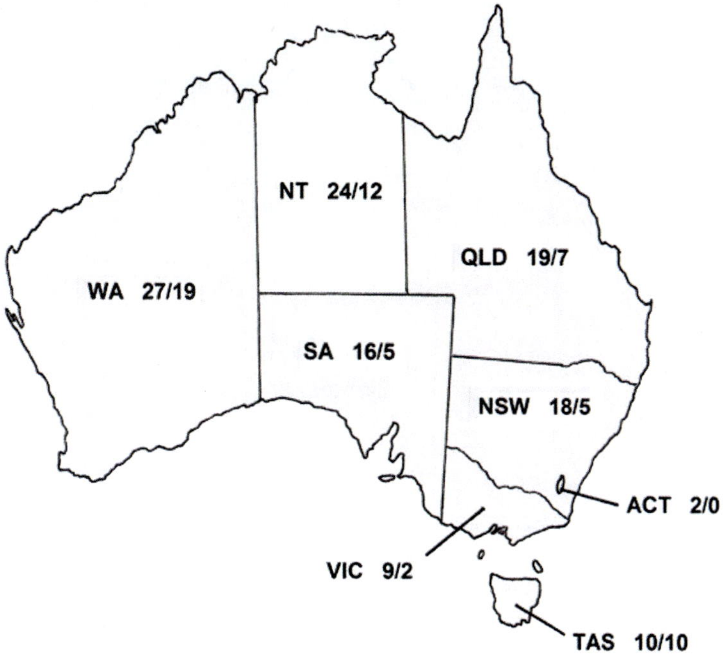

Fig. 7.1 The numbers of the Interim Biogeographical Regions (total 89) in different States and Territories of Australia, and the number of species of insects and allied taxa (namely, Entognatha, Arachnida, Myriapooda, Crustacea, Onychophora) listed as threatened in each region (if region or taxon transcends regional boundaries, accounted in both; shown as 'no.bioregions/no. taxa' (data from Taylor et al. 2018)

semi-arid regions, had no listed insect species. It is perhaps not surprising that most listed insects are from the more mesic areas, in which human populations and disturbances are largely concentrated—and also reflects that most entomological expertise and conservation awareness is situated in the more habitable southern and eastern parts of the continent. The 36 'vacant' bioregions reflect lack of conservation attention, and that the level of Data Deficiency among insects is high. Although insect species are indeed listed from the other 53 bioregions, numbers from many are very low and unlikely to represent the real needs of these sometimes complex regional faunas. Reducing this imbalance in insect representations was considered a priority for increasing the potential values of the NRS in those areas. Taylor et al. (2018) also noted that (1) five national biodiversity hotspots have very few designated threatened insect species and (2) several bioregions with higher numbers of listed species are not considered biodiversity hotspots. The overarching aim of the strategy focusing on flagship species is to ensure representativeness across all bioregions, with implications for increased management of all major habitats and ecological communities as well as key threatening processes. It is also expected that each such flagship insect will become an umbrella for numerous other species representing the wider diversity of those regions.

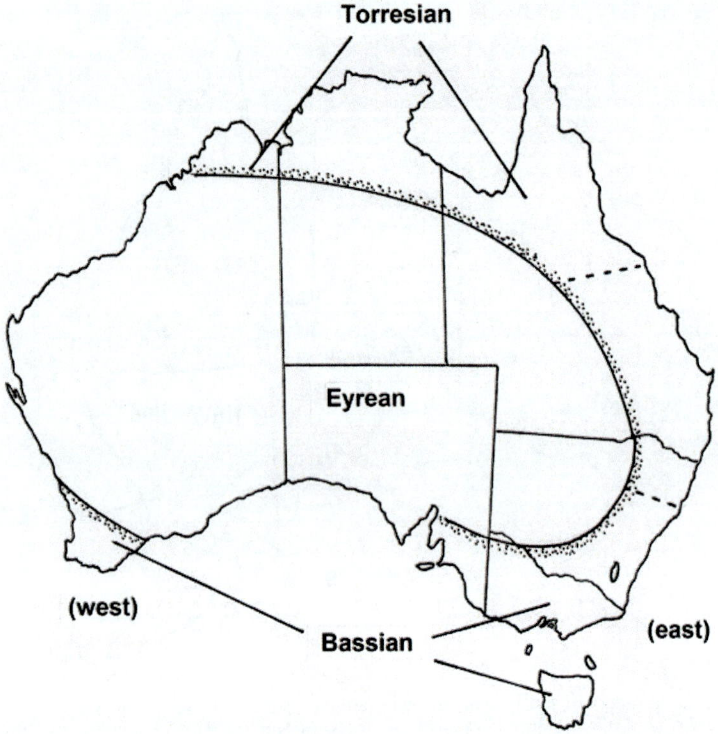

Fig. 7.2 The foundation of biogeographical regionalisation in Australia. The three major regions (Bassian/Torresian/Eyrean) distinguished initially by Spencer (1896) with later addition of the west Bassian province

7.3.2 Ecological Communities

As among their predominant biological components, insects can help to define the peculiarities and significance of ecological communities. Ability to list 'threatened ecological communities' under some Australian legislations is a vital part of planning for insect conservation from the basis of a few selected flagship taxa and extending perspective from species to wider entities. That most of the initially focal species are likely to have naturally restricted geographical and ecological distributions provides additional pressure for attention to key attributes of their critical habitat within such communities. Whether in the restricted higher elevation alpine/subalpine areas of the south east, the subterranean aquifers of the arid regions, the remnant patches of tropical and subtropical forest in the north east, or the immensely rich floristic assemblages of the south west, insect diversity is both peculiar and unexpectedly rich. These, and many other ecosystems, support insect taxa known nowhere else. Specific insect examples have already been listed or otherwise noticed for many ecological communities, augmenting other components of biodiversity in asserting conservation values, and an intention of the strategy

Fig. 7.3 The major terrestrial habitat categories in Australia, as reflected by predominant vegetation type: open, arid zone; vertical hatching, rainforest; horizontal hatching, sclerophyll forest; black, open woodland; dots, mallee

proposed by Taylor et al. (2018) is to extend that representation widely whilst also demonstrating the 'fine-grain' level of their requirements. There is also potential to nominate further threatened systems or areas for protection from the basis of their insect faunas.

Functionally, most threatened ecological communities have been characterised by physical features or attributes of their vegetation, and very few have specifically included insects or other invertebrates; these, as in much other conservation, are largely the passive beneficiaries of other priorities. For Australian butterflies, several restricted vegetation types and ecological communities were nominated for protection by Sands and New (2002), because each supported several endemic species that depended on those ecosystems and without them they were likely to become vulnerable.

'Ecological communities' are not the same as 'protected areas', but ambiguities in how these terms are interpreted by practitioners can lead to confusion. The 'essence' of a community is 'ecological' rather than 'spatial'. In Victoria, the entity listed as 'Butterfly community No 1' was recognised under the Flora and Fauna Guarantee Act to protect a unique suite of co-occurring resident and hilltopping Lepidoptera that included two very rare species of myrmecophilous Lycaenidae (Britton et al.

1995) that occurred at only one site in Victoria (Mount Piper, near Broadford). Any similar assemblage of Lepidoptera would fall under the Community ambit—but public and administrative perceptions of the entity listed confounded the community and place, so that 'Mt Piper' became synonymous with 'Butterfly community No 1', and the extent of differences in butterfly assemblage composition needed to adjudge any other assemblage sufficient to fall under this community or necessitate a further formal entity has not been established (New 2011).

7.3.3 Protected Areas

A threatened ecological community that falls within any protected area (such as a national park) may augment the formal need for protection and clearly falls under existing obligations for conservation. Protected areas have a range of roles and purposes, with a central one being conservation but accompanied by the realisation that biodiversity conservation 'is not and should not be confined to protected areas and that a significant proportion will remain outside protected areas' (Lopoukhine and de Souza Dias 2012). Modern insect conservation involves both protected areas with their often relatively undisturbed environments and also the vast areas of anthropogenic environments represented by agricultural, forestry and urban developments now major features of landscapes over much of Australia, and across which insect conservation must be integrated. As Samways (2020) emphasised, pursuing insect conservation in 'novel landscapes that are recrafted according to agroecological and urban greenspace principles' changes the more traditional perspective of relying more completely on protected areas, as a scenario that recognises (and seeks to redress) that most insect declines and losses have been caused by changing terrestrial and inland water ecosystems from natural to anthropogenic contexts. Nevertheless, protection of sensitive sites is vital in assuring that further disturbances are minimised and that management can avoid undue harm to target taxa and environments. More widely, some form of 'security of tenure' allows conservation investigations and management to proceed and be monitored. Protected areas form the basis of most national strategies for biodiversity conservation—and often in the case of insects and other unheralded taxa with the tacit assumption that simple protection of areas in which they occur will 'automatically' conserve them without further attention or management. In general, there is wide perception that the 'most natural' areas provide the greatest security and array of conservation options for native insect species, and that degradation of an area is likened to erosion of insect diversity—whilst also perhaps facilitating colonisation by others. Specialist species that depend on near-climax environments may be especially susceptible to disturbances—as Dolny et al. (2021) implied for European Odonata, for which degradation of notable habitats such as raised bogs and fens were associated with declines of already endangered specialist species, in circumstances that may apply widely elsewhere. In Australia, as in many other countries, many protected areas have been designated only recently and are of sites that have a long history of degradation and are far from their pristine state. Key conservation functions of protected areas are predominantly two-fold: (1) to maintain existing populations of the species present, so retaining their evolutionary viability as fully as possible, and (2) to reduce the risks of extinction. Criteria such as area, number and connectivity of protected areas across the wider landscape are clearly influential, with an ideal being a large number of bigger areas functionally connected as a network over the region. For any designation of protected areas as conservation reserves, harmonising different social priorities demands a balance between maximising the number of species represented and minimising the area of land sacrificed from other roles or uses. The principle of 'minimal reserve set' (Cabeza and Moilanen 2001) reflects that many reserves are primarily set up for political and/or economic reasons rather than on scientific criteria. In practice, many reserves have been designated without use of explicit criteria, and applications of the various computational methods devised to help identify sets of reserves that can maximise representation of biodiversity have been very inconsistent. Many selection criteria, when stated, are founded in principles of high diversity (nominally 'hotspots', as the most common single approach), complementarity, and representativeness within networks. However, at any scale of reserve size practicality dictates that the conservation ideals are tempered by social, economic and political constraints in their designation and design (Kati et al. 2004), even when scientific evaluation is compelling. As Kati et al. put it, 'in practice, conservation decisions are informed, not dictated by, science'.

Combination of long-term biodiversity conservation functions with conservation of ecosystem services and cultural values is implicit in the IUCN definition of 'protected area' (Dudley 2008; Dudley et al. 2013), and which is embraced in the various levels of protection recognised in the different area categories (Table 7.1). As AICHI Target 11, a global protected area network is planned to be 'ecologically representative', but does not specify how 'representativeness' is to be determined. However, Woodley et al. (2012) advocated some form of 'ecoregions' as a basis for this, with ecosystems viewed as part of delimiting areas for species conservation and for which connectivity is critical to prevent isolation. An increasing concern is whether the distribution and connectivity of protected areas can cater adequately for biological changes flowing from climate change. Another widespread consequence of having even a reasonably adequate protected area system is that all land outside those areas can then be considered 'expendable' and fit for development or exploitation without regard to other values. Wilson's (2016) plea that the extent of natural areas should be increased to include half the word's surface—although unlikely to be achieved—echoes the belief of many scientists that protected areas are indeed vital for conservation of Earth's biodiversity and that the current allocation is manifestly inadequate to achieve this. Representation of insects and other 'unseen diversity' in protected areas is manifestly inadequate, as inferred from species distribution models of selected Lepidoptera, Coleoptera and Hymenoptera for Costa Rica, Mexico and the United States (Delso et al. 2021), in all of which considerably increased area would be needed to increase representation.

As Sands (2018) noted, although many Australian insects have been found only in declared protected areas such as designated national parks, most of those areas

Table 7.1 The World Conservation Union (IUCN) categories of protected areas (IUCN 1994)

Category Ia: Strict nature reserve Strictly protected area set aside to protect biodiversity and also possibly geological/geomorphological features, where human visitation, use and impacts are strictly controlled and limited to ensure protection of the conservation values. Such protected areas can serve as indispensable reference areas for scientific research and monitoring.
Category Ib: Wilderness area Protected areas are usually large unmodified or slightly modified areas, retaining their natural character and influence, without permanent or significant human habitation, which are protected and managed so as to preserve their natural condition
Category II: National park Protected areas are large natural or near-natural areas set aside to protect large-scale ecological processes, along with the complement of species and ecosystems of the area, which provide a foundation for environmentally and culturally compatible spiritual, scientific, educational, recreational and visitor opportunities
Category III: Natural monument or feature Protected areas are set aside to protect a specific natural monument, which can be a landform, sea mount, submarine cavern, geological feature such as a cave or even a living feature such as an ancient grove. They are generally quite small protected areas and often have high visitor value
Category IV: Habitat/species management area Protected areas aim to protect particular species or habitats and management reflects this priority. Many category IV protected areas will need regular, active interventions to address the requirements of particular species or to maintain habitats, but this is not a requirement of the category
Category V: Protected landscape/seascape A protected area where the interaction of people and nature over time has produced an area of distinct character with significant ecological, biological, cultural and scenic value and where safeguarding integrity of this interaction is vital to protecting and sustaining the area and its associated nature conservation and other values
Category VI: Protected area with sustainable use of natural resources Protected areas are generally large, with much of the area in a more-or-less natural condition and where a proportion is under sustainable natural resource management and where low-level non-industrial use of natural resources compatible with nature conservation is seen as one of the main aims of the area

have not been comprehensively surveyed to determine what insects live there and, thus, their values in insect conservation. Those surveys are urgently needed (Sands and New 2003), not least because populations of significant insect species in high level reserves have potential for basic study and informed habitat conservation in an environment of secure tenure and freedom from unplanned deliberate interference. Even for the most high-profile national parks in Australia, insect inventories, other than for (in some cases) the most popular taxa, are highly inadequate. Some are undoubted centres of insect richness and endemism—and that awareness sometimes flows largely from historical collector lore. However, for many parks, including the major park of the Northern Territory, Kakadu, information is restricted and/or biased. Andersen et al. (2014) expressed that 'the vast majority of invertebrate species in Kakadu are poorly known even by specialist scientists, and it is likely that a substantial proportion has not yet even been collected' so that, except for some common and conspicuous species, 'there is simply insufficient information for assessing their conservation status'. For Kakadu, the greatest knowledge gaps are for terrestrial fauna, because many of the aquatic macroinvertebrates have indeed been surveyed more comprehensively with the main purpose of environmental monitoring in relation to mining activities. Consequently, the local dragonflies and some other orders are reasonably well documented, and 'uranium mining and tourism' is listed as a threat to three species categorised as 'Vulnerable' (listed by Andersen et al. 2014). Contrasts between effectiveness of protected areas for terrestrial and freshwater biodiversity are sometimes evident, although both are again embraced fully within the hierarchy of protected areas designated by IUCN (1994, Table 7.1).

Most protected areas have focused on terrestrial ecosystems, whilst many significant freshwater areas are protected only 'by default' as being included in terrestrial areas, where their key needs are often not attended specifically. Those needs (Saunders et al. 2002) focus on whole catchment integrity, hydrology, and introductions of non-native species. Thus, it is not uncommon to introduce non-native fish into protected areas for recreational angling, and Saunders et al. emphasised the need for reserves that specifically target freshwater habitats and protect biodiversity, representative habitats, rare or endangered species and intact habitat remnants. Many such reserves are subject to impacts from land use and hydrological changes beyond their boundaries. Ideally, freshwater protected areas should be in intact catchments, have natural hydrological regimes and lack non-native species. Their integrity depends on adopting strategies to address disruptions from both inside and outside the reserves. Discussions of how to increase attention to freshwater protected areas (Abell et al. 2007) recognise that they may need to be treated differently from the more predominant terrestrial protected areas, perhaps most notably in acknowledging that many freshwater systems are not circumscribed as easily as many terrestrial ecosystems, so that 'place-based protection' for aquatic systems extending far beyond that place (such as catchments or flowing rivers) and subject to numerous inputs from enveloping terrestrial environments may be unsuccessful. Many protected areas founded in terrestrial ecosystems contain waterbodies that have traditionally been largely disregarded: in Abell et al.'s words, freshwaters in many protected areas are 'buried within terrestrial accounting sheets and rarely recognised as their own special case'. Aquatic environments and species are often included fortuitously in protected areas: perhaps only within the Ramsar framework are specific aspects of protection of freshwater biodiversity sufficiently addressed.

The historical intention of Australia's national parks, quoted from Morcombe (1974), was 'to preserve, intact, segments of the natural environment' and to 'preserve land, plant and animal life in the balanced relationship' but, as Sands (2018) emphasised, the broader recent uses of national parks for sport, commercial or recreational activities, together with inadequate budgets for conservation management, and devolution of management to more local levels have compromised those ideals. Sands noted that environmentally insensitive activities in national parks are now permitted by some state agencies, citing examples in Queensland in which livestock grazing, horse and motor bike riding and other activities have been allowed.

Changes in protected areas that lessen protection are threefold: downgrading of status, downsizing, and degazettal (so summarised as 'PADDD'), together reflecting the difficulties of continuing to rely on any network of effective protected areas. Implications for Australia, discussed by Cook et al. (2017), involved the balance of considerable overall increase in total designated protected area and more than 1500 PADDD events that affected more than a third of the overall network; in all, 5233 changes in area and level of protection occurred over 17 years (1997–2014). One hundred and eight protected areas were degazetted, many of them small. The 1811 protected areas added were also mostly small, but 11 were classed as 'very large' ($>$10,000 km^2). The overall outcome implied that any 'permanent' protection for biodiversity remained somewhat uncertain. 'Downgrading' was linked commonly with infrastructure and constructions to promote tourism within protected areas, reflecting perceived changes in public appreciation toward utilitarian values and lessening emphasis on biodiversity conservation. The trend is also reflected in increased natural resource extraction activities. In order to assure continued protection of biodiversity, all PADDD events must be rendered transparent and their implications clearly understood and managed (Cook et al. 2017).

Protected areas can be effective in conserving biodiversity only if they are properly managed and resourced. Many are not, or any conservation attention does not actively consider insect diversity and representation. Across all taxa, 'many threatened species are afforded little or no protection within Australia's protected area network' (Kearney et al. 2019). As noted above, many protected areas have been opened up to activities that are not compatible with conservation: logging of old-growth forests is a notable example (Ritchie et al. 2013). Loss of forests, with Australia among the 11 global hotspots for deforestation (WWF 2015) is a major concern. Wider assessment of Australia's forests and woodlands demonstrated that 7.7 million hectares of potential habitat for terrestrial threatened or migratory species and threatened ecological communities was cleared over 2000–2017, and that a very high proportion of this loss (7.7millon ha, 93%) was not evaluated under the EPBC Act (Ward et al. 2019). This leading flagship legislation was thereby regarded as 'almost completely ineffective at limiting the ongoing loss of potential habitat' for these terrestrial taxa and communities. A recent comprehensive review of the Act (Samuels 2020) revealed many shortcomings in its operation and recommended substantial changes in order to restore credibility and render the Act fit for purpose.

Harmonising conservation with wider public use and amenity and economic values in such areas demands very fine balance in management and maintaining community support, but with primary need to preserve those areas in which notable insect taxa or communities occur.

The values of national parks as sanctuaries for Australian insects are appreciated widely by entomologists, but records of the insects present are most commonly serendipitous rather than the outcome of targeted surveys, and are taxonomically biased. Thus, a website listing (Department of Environment and Science 2013) for Lamington National Park, Queensland, comprises 42 butterflies and one damselfly.

Nevertheless, some national parks are famed for either their insect diversity or for harbouring or potentially harbouring notable endemic taxa of signalled conservation importance. As contrasting examples of current significance: (1) the entire range of the locally endemic satyrine butterfly *Oreixenica latialis theddora* (Alpine silver xenica) is in the Mt. Buffalo National Park, Victoria, where it is widely distributed on grasslands of this isolated plateau (Dunn 2019)—within the park a Special Protection Zone is an added management overlay to increase protection of the butterfly's major habitat area (Marion 1998); (2) headwater streams in the Alpine National Park, Victoria support notable endemic stoneflies (*Eusthenia* ssp.), and several other notable aquatic insects are associated with alpine lakes in the Kosciuszko National Park, New South Wales; and (3) Iron Range National Park, Queensland, is renowned for its high butterfly diversity, with some species not known to breed elsewhere. There is little to imply that parallels do not occur for many other insect orders across a varied range of parks, and would be revealed by more extensive inventory surveys. Encouragement of such surveys, and their facilitation by responsible issues of permits and responsible treatment of the data and specimens accumulated would augment knowledge of their roles considerably, together with increased awareness of threats to resident species. Protected areas also provide opportunities for future conservation, such as for re-introductions of the Golden carpenter bee (*Xylocopa aerata*, p. 204) into the Grampians National Park, Victoria as a response to changing site suitability as climate changes occur. There is also little doubt that the insect fauna of many national parks contrasts markedly with the more restricted and changed fauna beyond their borders and are likely to contain taxa extirpated from the wider enveloping landscape. The overarching priority need for insect conservation over vast areas of underexplored tropics, and illustrated effectively from Janzen's long-term efforts for the Area de Conservacion Guanacaste (Costa Rica, p. 77), is to conserve and restore 'very large and habitat-diverse and ecosystem-diverse wild areas', with old-growth forest signalled as especially important in the tropics.

Protected area needs extend well beyond the concept of national parks, which are largely an 'umbrella' for predominantly undocumented insect diversity, with the presumption that many of the species there do not occur, or are more vulnerable, outside those parks. The same applies to some other high-profile areas, notably those designated as 'biodiversity hotspots' characterised by high biodiversity, and often with many endemic components. Protection of hotspots has been advocated as one of the best ways to conserve a large proportion of biodiversity, but is not on its own sufficient as that reliance may lead to neglect of other critical environments (Marchese 2015). As Marchese put it 'biodiversity is much more than just the numbers of species in a region'. Diversity and richness are key criteria in setting priorities, but many small protected areas are important for harbouring highly significant taxa, such as localised archaic or taxonomically isolated species rather than high richness alone. However, reliance on a protected area for full conservation can be short-sighted. The small colletid bee *Neopasiphae simplicior* is known from a single population in a nature reserve near Perth, Western Australia, where its total range is around 1 km^2. It has been listed as critically endangered under EPBC, reflecting a range of possible threats to that reserve. However, the EPBC Advisory Committee considered that a recovery plan for *N. simplicior* was unnecessary because it was known wholly from a managed reserve and information on threats was deemed sufficient to manage them. Advice offered to the Minister for consideration had included several strong recommendations to monitor the bee and protect it within the reserve.

Because some insects can sustain viable populations on areas of only a few hundred m^2, very small reserves (below the size levels that would often be considered useful for other biota) may become vital, with due regard to erosion of habitat integrity through the increased edge effects that could increase disturbance from outside. However, the ideal of a complete insect inventory for any Australian national park or other designated reserve is far off. Pringle's (2017) comment that 'A protected area without an inventory of its resident species is like a library without a catalogue, a supermarket of unlabelled cans, and a museum with all of its artworks piled in the basement' expresses the frustrations and lack of credibility when advocating for their conservation on grounds of 'biodiversity value'. Lack of knowledge both of notable individual species being present and overall richness of the insect fauna can seriously erode arguments for conservation values.

The very term 'hotspot' is interpreted in several different ways—Braby et al. (2020, following Marchese 2015) listed three main interpretations as: (1) areas that support high levels of endemic species, so that habitat loss would threaten many of these highly localised taxa; (2) areas with exceptionally high species richness, with many taxa likely to be range-restricted; and (3) areas that support high levels of phylogenetic diversity, phylogenetic endemism or functional diversity. However, hotspots are largely indicative of where conservation priorities might be allocated, rather than protected per se, and most such areas are already severely degraded. Historically, these areas have been erected on levels of richness and endemism of plants and vertebrates without consideration of invertebrates—so that the demonstrated considerable significance of the former groups is accompanied by considerable ignorance of others. Stork and Habel (2014) suggested that the potential role of plants as 'umbrella species' for numerous herbivorous insects and their natural enemies may confer surrogate values, but that these roles need to be explored in far more detail. It has been reported frequently that hotspots of diversity based on any given taxonomic group may only rarely coincide with those defined from other taxa—so that networks of protected areas needed for different taxa may differ (Kati et al. 2004). Hotspots include large areas over which natural habitats have been lost, and where considerable resident human populations continue to influence local environments. Proportions of formally protected natural environments there are low, threats are largely unabated, and insect life is woefully underdocumented. The 'Forests of East Australia', designated as the world's 35th Biodiversity Hotspot (Williams et al. 2011), comprises the Tropical Rain Forests of Queensland and the wide spread of coastal/subcoastal temperate forests further south (Fig. 7.4), together containing more than 1500 endemic plants and areas in which more than 70% of natural vegetation has been cleared or degraded. Only about 18% of the total land area is under any form of formal protection that reflects its natural values.

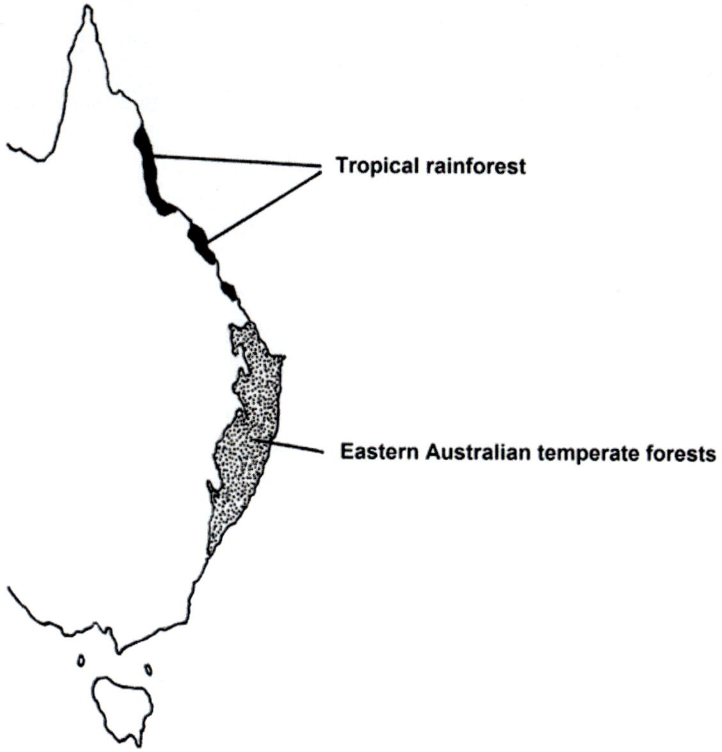

Fig. 7.4 The Forests of East Australia hotspot: distribution of the two ecoregion components as Queensland Tropical Rain Forests (north) and Eastern Australia Temperate Forests (south) (based on Williams et al. 2011)

In practice, conservation focus is necessarily on more local 'hotspots within a hotspot', and within any such more local entity, 'micro-hotspots' constitute smaller areas in which focused management may be most practicable and beneficial. Grant and Samways (2011) defined micro-hotspots as 'overlapping areas of endemism and rare species that are under threat at the landscape scale'. Within these, specialised local endemic species are of high concern. Identification of such areas in larger biosphere or other reserves—in Grant and Samways' study, for Odonata in the Kogelberg Biosphere Reserve in the Cape Floristic Region of South Africa, a designated global hotspot—has critical importance for reserve management, in this case for removal of alien vegetation and monitoring the threatened endemic dragonfly assemblage. Protected or managed buffer zones around such micro-hotspots can augment targeted management. Much practical insect conservation takes place at small spatial scales, whilst the bulk of higher-level conservation planning is for larger scales, and is based on relatively coarse biological information, as discussed by Cabeza et al. (2010).

Other remnant areas of natural vegetation or aquatic ecosystems also reflect that many insects can persist on small areas that would often be overlooked or disregarded in management and which in many cases could become functionally linked by 'stepping stones' or more continuous corridors. Small remnant forest or native grassland patches are both important in Australia. Both protection and restoration of such areas may be practicable, and the latter becomes critical in enhancing values of degraded remnants such as in urban and periurban areas. Recent appraisals of importance of urban areas for insects (New 2015, 2018) and the values of roadside habitats (New et al. 2021) embrace many sensitive and significant environments under the administrative control of local government bodies such as local councils and not necessarily signalled at higher levels as important. Some such areas are indeed important for particular insects: the stronghold of the Bulloak jewel butterfly (*Hypochrysops piceatus*, Lycaenidae) (p. 203) in Queensland is on a short stretch of roadside that has been declared as a Nature Refuge (Ellangowan Nature Refuge), and clearing of mallee vegetation along highways in South Australia was regarded as a threat to the Dinosaur ant (*Nothomyrmecia macrops*) (Watts et al. 1998). There are abundant opportunities to extend that approach of acknowledging significance of small roadside vegetation fragments for insects. Recognition of the botanical significance of roadsides is more advanced, with many roadside lengths signalled across the continent to protect them from management measures such as mowing or slashing. Many of those areas are also likely to foster local associated insects. In an idealistic principle, any area that supports a unique or highly restricted insect, plant or other biotic component may merit protection.

Assessing the reality of protection provided by a designated protected area demands careful investigation. Three functional categories of effectiveness of protected areas for focal species were defined for South Africa (Anon 2017), and are a useful model for emulation in Australia. These categories are (1) poor: protected area provides no mitigation of major threats to the species, so that individuals inside the protected area are no better off than those outside; (2) fair: protected area provides some mitigation of major threats to the species, but is not 100% effective; and (3) good: protected area is fully effective in protecting the species against major threats and ensuring the long-term persistence of the population present within the protected area. This scheme was applied to conservation status assessments of South African butterflies, acknowledging that if a threatened insect is known to occur in several protected areas it may be less threatened than if it occurred in fewer, or none at all. Mecenero et al. (2021) used this as a component of 'protection level', based on the extensive general distribution knowledge of that butterfly fauna accumulated through targeted surveys. Thus, the South African protected area network comprised 1273 areas of national parks, provincial and private nature reserves, and others. Butterfly taxa occurring in >10 protected areas were assessed as 'well protected', pending investigations to determine if further evaluation was needed—for example, if only very small populations were present. For species in 10 or fewer protected areas, in many cases reflecting local endemism, estimates of protection effectiveness (above) were sought, in many cases leading to an 'unknown' outcome because of lack of knowledge of that particular area, and further consensus might be sought from informed lepidopterists. Further evaluation criteria described by Mecenero et al. enlarged on this approach to evaluating effectiveness of protected areas, in conjunction with mapping of critical habitat for each threatened species. Each taxon in 10 or fewer protected areas was given a calculated 'protection level' to reflect the total number of subpopulations and allocated to one of four categories in relation to representation and likely viability as 'not protected', 'poorly protected', 'moderately protected' or 'well protected'. The additional category of 'unknown' was again applied to taxa for which information was insufficient to categorise fully.

However, whatever formal protected status is given, simply 'locking up' an area and designating it as 'protected' is only a first step—but an invaluable one—in practical conservation. Essentially, it contributes to slowing or reducing many of the more obvious threats from destruction and intrusion and provides opportunity for investigation and focused conservation management in a more secure environment, where focus can be refined and a longer-term vision come to the fore. It cannot be assumed that the biota contained in any protected area are indefinitely (or even temporarily) 'safe', even if that biota is documented sufficiently to assess risks to its security. Biological information on the insects (and many other taxa) is particularly lacking from many protected areas in the tropics, and is highly incomplete for most temperate region reserves. But, as Klorvuttimontara et al. (2011) commented, it is often unclear whether tropical protected areas are well-placed for conserving biodiversity either at present or under changed climatic conditions in the future. There is more general sentiment that the global extent of protected areas is still vastly inadequate to prevent continuing loss of biodiversity and Pringle (2017) and others have emphasised the urgent need to, in his words, 'enlarge the protected area estate, to increase its overlap with endangered ecosystems and at-risk species, and to reconcile the areas with projected land use changes' as human populations increase. Many calls to enlarge the protected area system are impracticable because of competing interests for land use, and also because landscape disturbance has already been sufficiently extensive that optimal candidate areas are severely degraded and the extent of losses already sustained of insects and other biota are unknown. Small areas dedicated for conservation of individual species may in reality be only marginal in providing the critical resources needed, and in allowing connectivity with other such areas.

With adequate determination and support, surveys could go far to censusing the current insect fauna present (at least in, perhaps overly simplistic, morphological species enumeration) but assessing future suitability of a reserve is still highly uncertain. Conservation values could increase with influx of other species driven from elsewhere by climate change, but be balanced by the area becoming unsuitable for some current residents, leading to mortality, abandonment or changed interactions from which ecological specialist insects of greatest conservation priority might prove especially vulnerable. Most interactions may be affected, some in unanticipated ways. Moir (2015) for examples, exemplified (1) changed impacts from natural enemies and (2) that plant populations regenerating after fire may have lost their specialist herbivores. Consideration of these and other unpredicted responses is largely impracticable in any part of the world.

References

Abell R, Allan JD, Lehner B (2007) Unlocking the potential of protected areas for freshwaters. Biol Conserv 14:48–63

Andersen AN, Humphrey C, Braby MF (2014) Threatened invertebrates in Kakadu National Park. In: Widerlich S, Woinarski J (eds) Symposium 7: conservation of threatened species, 26–27 March 2013, internal report 623. Supervising Scientist, Darwin, pp 48–57

Anon (2017) Setting targets and measuring protection levels for species: a proposed method. South African National Biodiversity Institute, Pretoria

Australian Government (2017) Australia's bioregions (IBRA). Department of Agriculture, Water and the Environment, Canberra

Braby MF, Williams MR, Coppen RAM, Williams AAE, Franklin DC (2020) Patterns of species richness and endemism of butterflies and day-flying moths in the monsoon tropics of northern Australia. Biol Conserv 241:108357

Britton DR, New TR, Jelinek A (1995) Rare Lepidoptera at Mount Piper, Victoria - the role of a threatened butterfly community in advancing understanding of insect conservation. J Lepidopt Soc 49:97–113

Cabeza M, Moilanen A (2001) Design of reserve networks and the persistence of biodiversity. Trends Ecol Evol 16:242–248

Cabeza M, Arponen A, Jaattela L, Kujala H, van Teeffelen A, Hanski I (2010) Conservation planning with insects at three different spatial scales. Ecography 33:54–63

Cook CN, Valkan RS, Mascia MB, McGeoch MA (2017) Quantifying the extent of protected-area downgrading, downsizing, and degazettement in Australia. Conserv Biol 31:1039–1052

Delso A, Fajardo J, Munoz J (2021) Protected area networks do not represent unseen biodiversity. Sci Rep 23:12275. https://doi.org/10.1038/s41598-021-91651-z

Department of Environment and Science, Queensland (2013) Insects of Lamington National Park. Wetland Info website. Accessed 8 June 2021

Dolny A, Ozana S, Burda M, Harabis F (2021) Effects of landscape patterns and their changes to species richness, species composition, and the conservation value of odonates (Insecta). Insects 12:478. https://doi.org/10.3390/insects12060478

Dudley N (ed) (2008) Guidelines for applying protected area management categories. IUCN, Gland

Dudley N, Shadie P, Stotton S (2013) Guidelines for applying protected area management categories including IUCN WCPA best practice guidelines on recognising protected areas and assigning management categories and governance types. Monographic series no 21. IUCN, Gland

Dunn KL (2019) *Oreixenica latialis* (Lepidoptera: Nymphalidae: Satyrinae) on Mount Buffalo. Victoria: field notes on its local distribution and adult behaviour to assist future conservation work. Calodema 724:1–24

Ebach MC, Murphy DJ (2020) Carving up Australia's arid zone: a review of the bioregionalisation of the Eremaean and Eyrean biogeographic regions. Aust J Bot 68:229–244

Farrier D, Whelan R, Mooney C (2007) Threatened species listing as a trigger for conservation action. Environ Sci Pol 10:219–229

Grant PBC, Samways MJ (2011) Micro-hotspot determination and buffer zone value for Odonata in a globally significant biosphere reserve. Biol Conserv 144:772–781

Hutchings P (2004) Invertebrates and threatened species legislation. In: Hutchings P, Lunney D, Dickman C (eds) Threatened species legislation: is it just an act? Royal Zoological Society of New South Wales, Mosman, pp 88–93

IUCN (1994) Guidelines for protected areas management categories. IUCN, Gland and Cambridge

Kati V, Deviers P, Dufresne M, Legakis A, Vokou D, Lebrun P (2004) Hotspots, complementarity or representativeness? Designing optimal small-scale reserves for biodiversity conservation. Biol Conserv 120:471–480

Kearney SG, Carwardine J, Reside AE, Fisher DO, Marin M et al (2019) The threats to Australia's imperilled species and implications for a national conservation response. Pac Conserv Biol 25: 231–244
Kitching RL, Dunn K (1999) The biogeography of Australian butterflies. In: Kitching RL, Scheermeyer E, Jones RE, Pierce NE (eds) Biology of Australian butterflies, Monographs on Australian Lepidoptera, vol 6. CSIRO Publishing, Collingwood, pp 53–74
Klorvuttimontara S, McLean CJ, Hill JK (2011) Evaluating the effectiveness of protected areas for conserving tropical forest butterflies of Thailand. Biol Conserv 144:2534–2540
Lopoukhine N, de Souza Dias BF (2012) Editorial: what does target 11 really mean? Parks 18:5–8
Marchese C (2015) Biodiversity hotspots: a shortcut for a more complicated concept. Glob Ecol Conserv 3:297–309
Marion A (1998) Special protection zone management within Mount Buffalo National Park. Vict Nat 115:169–174
Mecenero S, Edge DA, Staudte HS, Coetzer BH, Coetzer AJ et al (2021) Outcomes of the southern African Lepidoptera conservation assessment (SALCA). Metamorphosis 31:1–160
Moir M (2015) Conservation of Australia's plant-dwelling invertebrates in a changing climate. In: Palutikof JP, Boulter SL, Barnett J, Rissik D (eds) Applied studies in climate adaptation. Wiley, Chichester, pp 107–116
Morcombe M (1974) Australia's national parks. Lansdowne Press, Melbourne
New TR (2011) Butterfly conservation in South-Eastern Australia: progress and prospects. Springer, Dordrecht
New TR (2015) Insect conservation and urban environments. Springer, Cham
New TR (2018) Promoting and developing insect conservation in Australia's urban environments. Austral Entomol 57:182–193
New TR, Sands DPA, Taylor GS (2021) Roles of roadside vegetation in insect conservation in Australia. Austral Entomol 60:128–137
Pringle RM (2017) Upgrading protected areas to conserve wild biodiversity. Nature 546:91–99
Ritchie EG, Bradshaw CJA, Dickman CR, Hobbs R, Johnson CN et al (2013) Continent-scale governance and the hastening of loss of Australia's biodiversity. Conserv Biol 27:1133–1135
Samuels G (2020) Independent review of the EPBC act – interim report. Department of Agriculture, Water and the Environment, Canberra
Samways MJ (2020) Insect conservation. A global synthesis. CABI, Wallingford
Sands DPA (2018) Important issues facing insect conservation in Australia: now and into the future. Aust Entomol 57:150–172
Sands DPA, New TR (2002) The action plan for Australian butterflies. Environment Australia, Canberra
Sands DPA, New TR (2003) Coordinated invertebrate surveys in Australia's National Parks: an important tool in refining invertebrate conservation management. Rec South Aust Mus Monogr Ser 7:203–207
Saunders DL, Meeuwig JJ, Vincent ACJ (2002) Freshwater protected areas: strategies for conservation. Conserv Biol 16:30–41
Spencer WB (1896) Report on the work of the Horn Scientific Expedition to Central Australia. Part 1. Introduction, narrative, summary of results, supplement to zoological report, map. Melville, Mullen and Slade, Melbourne
Stork NE, Habel JC (2014) Can biodiversity hotspots protect more than tropical forest plants and vertebrates? J Biogoeogr 41:421–428
Taylor GS, Braby MF, Moir ML, Harvey MS, Sands DPA et al (2018) Strategic national approach for improving the conservation management of insects and allied invertebrates in Australia. Austral Entomol 57:124–149
Thackway R, Cresswell ID (eds) (1995) An interim biogeographic regionalisation for Australia: a framework for setting priorities in the National Reserve System Cooperative Process. Version 4.0. Australian Nature Conservation Agency, Canberra

Walsh JC, Watson JEM, Bottrill MC, Joseph LN, Possingham HP (2013) Trends and biases in the listing and recovery planning for threatened species: an Australian case study. Oryx 47:134–143

Ward MS, Simmonds JS, Reside AE, Watson JEM, Rhodes JR et al (2019) Lots of loss with little scrutiny: the attrition of habitat critical for threatened species in Australia. Conserv Sci Pract 2019(1):e117

Watts CHS, McArthur A, Foster R (1998) Notes on the distribution of the dinosaur ant *Nothomyrmecia macrops* Clark (Hymenoptera: Formicidae) in South Australia. Aust Entomol 25:29–31

Williams KJ, Ford A, Rosauer DF, De Silva N, Mittermeier R et al (2011) Forests of East Australia: the 35th biodiversity hotspot. In: Zachos FE, Habel JC (eds) Biodiversity hotspots. Springer, Berlin and Heidelberg, pp 295–310

Wilson EO (2016) Half-earth: our planet's fight for life. Norton, New York

Woodley S, Bertzky B, Crawhall N, Dudley N, Londono JM et al (2012) Meeting Aichi target 11: what does success look like for protected area systems? Parks 18:234–235

WWF (World Wide Fund for Nature) (2015) Living Forest Report: saving forests at risk. World Wide Fund for Nature, Gland

Chapter 8
Defining and Countering Threats

8.1 Introduction

Anthropogenic disturbances to the Australian environment, assessed generally as only those occurring since European settlement and superimposed only from the mid eighteenth century on a far longer history of traditional management by First Nations peoples, are relatively recent when compared with those in much of the rest of the world. They have become widespread, varied and intensive, with transformation of much of Australia to conditions far different from those of two centuries ago. The concentration of human populations toward the coastal eastern and southern regions, largely within the more clement temperate regions and with historical population centres (including all major capital cities) widely separated, has been accompanied by large-scale agricultural and pastoral activity—with some single grazing properties in inland Australia of size equivalent to some European countries—and exploitation of the country's rich coal and mineral ore deposits as a major component of export industry. Nevertheless, until quite recently human pressures on the more inaccessible or remote parts of Australia have been relatively low and these have remained undisturbed and nominally 'pristine', together with their resident biota. In principle, those areas have potential to act as 'reference sites' that can furnish some form of baseline to more closely characterise the pre-European insect fauna and evaluate the extent of faunal changes and their causes over the intervening two centuries. Those opportunities decrease as impacts continue to proliferate and extend far beyond their points of introduction, and the conservation values of the remaining undisturbed remnant environments and protected areas increase accordingly.

Threat evaluation can only rarely become quantitative for insects. The World Conservation Union Threat Impact Scoring System (IUCN 2017) includes consideration of timing (solely in the past; past but likely to re-occur; ongoing; only in the future; unknown), scope (affects whole population; >90%; affecting majority of the population, 50–90%; affecting minority of the population, <50%; unknown), and severity (posing a series of assessments of extent and rapidity of declines based on

© The Author(s), under exclusive license to Springer Nature Switzerland AG 2022
T. R. New, *Insect Diversity, Declines and Conservation in Australia*, Fascinating Life Sciences, https://doi.org/10.1007/978-3-030-90134-9_8

quantitative thresholds; limited declines; unlikely to cause declines; unknown). The quantitative thresholds, in principle paralleling those used for species conservation status assessments are largely impossible to derive for insects, but in a recent global assessment, only 65 of 1795 designated threatened taxa were invertebrates (Ward et al. 2021). The most severe threats to these, each affecting about 20% of the taxa, were cited as adverse fire regimes, invasive species and disease, and habitat loss and associated fragmentation and degradation.

'Australia is distinct not only in the biodiversity it contains but also in the extent and mixture of pressures that threaten the survival of those species' (Kearney et al. 2019). All the major 'drivers' of decline or threats to insects noted in Chap. 5 apply in Australia, some very severely—but, in general, without heed or knowledge of their impacts on native invertebrates, so that inferences on insect losses can come largely or only from assessing the extent of habitat loss, studying the species present in remnant habitat patches and presuming that these may indeed have been more widespread in past times. Within the great variety of potential threats, it is believed widely that destruction of native vegetation from the range of activities spanning agriculture, extraction and plantation forestry, mining, industrial and urban development, is the most significant threat to Australian insects. It has been claimed that since European settlement some 33% of all native vegetation has been lost (Ward et al. 2019), mostly in agricultural and coastal areas. 'Urban sprawl' continues to absorb natural vegetation, and several of Australia's major cities are located in or near recognised biodiversity hotspots that support significant insect species—which may become vulnerable as urbanisation proceeds (New 2018). Even significant vegetation associations, including those that represent legally declared threatened communities, are not immune from destruction, as such formal protection may lead people to believe. Elsewhere, few controls on vegetation losses have been implemented, with the Commonwealth EPBC Act still not filtering considerable losses of potential forest and woodland habitats since it was activated in 2000, leading Ward et al. (2019) to conclude that the Act was ineffective for protecting terrestrial threatened species, migratory species or threatened ecological communities. They also noted that individual habitat losses may not be large—but that progressive small losses by 'many small cuts' are often just as destructive as more abrupt and noticeable losses. Across all taxa of animals and vascular plants, more than 60% of federally designated threatened species (1136 of 1862 species) are severely threatened by habitat loss. Recommendation for determining and designating 'critical habitat' for all threatened species, and ensuring that clear guidelines for assessing any potential imposed disturbance extends far beyond insects, and is complicated by the varying implications of different land tenures (Sands 2018). Many important entomological sites are on private lands, where their maintenance can be at the whim of the individual proprietors, and change rapidly if the property changes hands. However, habitat degradation has many causes, amongst which invasive animals and plants play major roles. The Commonwealth's advice supporting listing of the Bathurst copper (or Purple copper butterfly), *Paralucia spinifera* (p. 156), for example, included the adverse impacts of feral goats and pigs as well as of several invasive plant species, notably blackberry, *Rubus fruticosus*, and Scotch broom, *Cytisus scoparius* (TSSC 2016). More widely, grazing and trampling by domestic stock on sensitive vegetation can be destructive—but, conversely, controlled/regulated seasonal or low intensity grazing can be a significant management tool in grassland and other low vegetation, and help also to influence vegetation succession. Maintenance of early successional stages of open vegetation is a common target of insect species conservation.

The drivers cited as affecting the greatest numbers of Australian threatened taxa are invasive species, ecosystem modifications and agriculture (Kearney et al. 2019), with invasive species the most widely cited threat across all taxa. Broad comparative assessments of threats for Australia's threatened animals and plants inevitably contain rather little specifically on insects, but Kearney et al. (2019) found ranking of threats to be reasonably consistent across taxonomic groups. Both invasive species and ecosystem modifications were important for greater proportions of Australian biota than for global threatened biota, with implication that attention to these components alone is key to effective conservation in Australia—and with insects major beneficiaries of such attention to other taxa. However, although some factors (such as urbanisation) were less cited for Australia than elsewhere, their continuing habitat transformations and impacts on highly localised taxa cannot be overlooked. As well as direct habitat and vegetation changes, 'ecosystem modifications' includes changes to fire regimes and hydrological alterations—so implicitly implicating climate change influences. Agriculture included livestock grazing as the dominant land use in Australia. Contrasts between ranking of key threats in Australia and globally notably included the reduced impacts of overexploitation and increased impacts of invasive species—the latter reflecting in part Australia's 'island status' with highly endemic biota perhaps especially susceptible to impacts of alien arrivals. However, Kearney et al.'s synopses agree with evaluations by Ward et al. (2021) that the primary need for conservation is mitigation of aspects of habitat loss, fragmentation and degradation, these driven largely by agriculture and urban development.

Many insect species are now known only from small remnants of particular vegetation associations or wetland forms. Habitat losses through land clearing for agriculture, pastoralisation and industrial and urban developments, together with changes to wetlands and inland waters of all kinds, have proceeded since European settlement, but accelerated markedly through the twentieth century, and with increasing concerns over impacts of climate change now superimposed on these. All have been addressed in detail, with concerns expressed for all components of Australian fauna and flora through both the spatial extent and severity of changes and removal of native vegetation. The potential impacts of widespread threatening processes have only rarely been tested empirically or by specific association with particular insect declines—other than in the more extreme contexts such as 'habitat destruction' from any of a variety of single or compounding causes, at extremes such as clearing of entire patches of forests or grassland or draining of wetland areas. Many of the more insidious and subtle effects of less extreme environmental changes on insects are difficult to quantify but may be implied strongly by extrapolation from parallel cases and by 'common sense' combined with precaution—neither of which may be politically convincing in conservation advocacy. However, insects were late arrivals in the Australian conservation literature (New and Yen 2012) and were largely passengers among the early emphases on vertebrates, for which vegetation integrity was also critical. More focused interest was stimulated in a classic essay by Marks (1969), as the first major call for invertebrate conservation in Australia and described as 'far ahead of its time' (New and Yen 2012). Recent concerns for insects have devolved largely on habitat losses, with notable attention also to impacts of alien species and fire, the latter reflecting the increased use of deliberate fire ('control burning' or 'fuel reduction burning') in the expectation of reducing intensity of wild fires and their danger to human life and property. Summaries of both these harbingers of environmental changes (New 2014, 2016) demonstrate their complexity and difficulties of predicting impacts on insect life. For both, those general perspectives imply their significant environmental impacts, and the concerns resonant in environmental policy and management. Alien species, together with climate change (p. 156), create opportunities for novel interactions between species that have perhaps never co-existed naturally; should the alien be an aggressive invader, native species may be vulnerable to effects of competition or attack in a newly defined community. Concerns for fire impacts must now include also the largely uncharted effects of 'megafires' linked to climate change. Following the major megafire over August 2019–February 2020, in which many unique forest ecosystems and threatened ecological communities were enveloped within the more than 10 million ha burned, reports estimated that 'hundreds of millions' of animals had perished, with many threatened species then closer to becoming extinct. Much of that inference was from vertebrates, and lack of baseline information largely precluded any similar interpretation for invertebrates (Saunders et al. 2021). Understanding impacts of wildfires on Australian insects, and designing policy to monitor and ameliorate these is in its infancy.

The more general impacts of 'habitat loss' from urban expansion (p. 150), together with establishment of supporting industrial and recreational needs and communication networks have led to widespread concerns for previously more natural biota and ecosystems affected directly or indirectly by those activities (New 2015), and emphasised the values of remnant natural patches within urban areas, especially when considered as components of a functional interconnected network. Losses of natural habitats as areas are subsumed to housing and related activities are obvious, but documenting declines of resident insects, other than in very general terms, is rare and reflects the paucity of reliable historical information sufficient for rigorous analysis. The butterflies of the region around Eltham, an outer north-eastern suburb of Melbourne, Victoria, were appraised (Braby et al. 2021) to investigate changes in richness and assemblage composition based on accumulated records of presence/absence over 40 and 80 years (1981–2020, 1941–2020), respectively. Three spatial scales were investigated to explore the fate of the 52 species present, of which 46 were resident or regular seasonal immigrants. Changes in all resident species occurred at all scales, and were most obvious at the smaller scales. Losses were most evident among ecologically specialised taxa, notably ant-associated lycaenids, and taxa whose larvae feed on native grasses or mistletoes, reflecting loss of specialised resources as urbanisation proceeded, habitat decrease and isolation eventuated and climate changes confounded direct changes by warming and drought. Ten of the 12 species considered extirpated still occur in the wider Eltham area.

Several key insect species conservation cases in Australia have developed directly from changes to periurban environments as housing and related developments continue to engulf natural ecosystems there. Less obvious disturbances can include changed storm water flows and pollution inputs (New 2015), either broadly or affecting restricted sites. A site supporting a small colony of the Eltham copper butterfly (p. 202) in central Victoria (Canzano and Whitfield 2008), for example, was inundated through increased storm water concentrated from newly cleared housing blocks on its periphery about 2013 (Threatened Species Scientific Committee 2016), and led to measures to reduce flow effects from future such events.

8.2 Pesticides and Pollution

Increased, and sometimes rather indiscriminate, use of agricultural and veterinary chemicals for crop and livestock protection, as well as a variety of domestic applications, has many implications for the wellbeing of ecological communities and their inhabitants. 'Run-off' and a wide range of industrial and urban pollutants reaching waterways are a major threat to aquatic insects (New 2021). Accusations and suggestions of non-target effects from pesticides and other pollutants are widespread in insect conservation, in parallel with those from classical biological control agents and in both contexts resulting from lack of specificity, and their facility to move or be moved into environments other than those for which they were intended. Pesticide drift to non-agricultural lands, or leaching or runoff into waterways has commonly been postulated. The amounts and variety of pesticides and other pollutants released to the natural environment are difficult to conjecture. As examples, in noting that pesticide use is better documented in the United Kingdom than elsewhere, Goulson (2019) cited 2015 government statistics that UK farmers applied a total of 16.9 thousand tons of active ingredient in an array of 17.4 different pesticides/hectare of land, leading to 'complex cocktails' of insecticides, fungicides and herbicides (including the neurotoxic neonicotinoids, below) in food for bees. He also noted the impacts of synthetic fertilisers and, extending beyond agriculture and Britain, that other pollutants were drawn from the approximately 144,000 manufactured chemicals, collectively comprising approximately 30 million tons and some of which have been found in virtually all environments examined.

Much recent attention has flowed from the uses of neonicotinoid insecticides as systemic pesticides intended for crop protection against insect pests and their toxic impacts on honeybees. These neurotoxic chemicals have been used on very large scales, and are highly persistent in soils, from where they can be transported easily into waterbodies.

High toxicity to honeybees has become a major concern, and has been reported repeatedly, together with implications for wider bee 'health' (Pisa et al. 2015), and commercial values have resulted in *Apis mellifera* being by far the most intensively studied non-target insect. Sublethal effects include implied synergistic effects with some key pathogens (such as *Nosema* spp.). Concerns have also been documented for other pollinators (including native bees), predatory insects (such as Carabidae) valued in crops and non-pest Lepidoptera (with needs for further studies on this order: Pisa et al.) and with pesticide transmission variously through contaminated plants, nectar, pollen or insect prey. Little or no information is available on the susceptibility of most insect groups in agricultural and neighbouring environments to these systemic insecticides. Although neonicotinoids (and other pesticides) undergo formal risk assessment before they are approved for release, the procedures have been described as 'inadequate' (Sgolastra et al. 2020) and in need of modification.

Toxicity to honeybees led the European Union in 2013 to 'severely limit' uses of neonicotinoids. Later (2018) they banned use of three major compounds (imadicloprid, clothianidin, thiamethoxcam) from all field uses, so confining them to greenhouse environments. In the meantime, widespread concerns from the earlier European Union ruling led to extensive reviews in many places, with additional calls for bans on their use. For Australia, a review (APVMA 2014) accepted that neonicotinoids are toxic to honeybees, but that this property was shared with many other insecticides. The review concluded (1) that Australian honeybees are not in decline, despite increased use of these chemicals and (2) that the chemicals had led to overall reduction in risks to the agricultural environment from insecticide applications. It also acknowledged that continued monitoring and research is needed to assess possible future changes. Progressive shifting of emphasis in Australian apiary from honey production to provision of pollination services is also likely to increase exposure of bees to insecticides. Concerns continue globally over the impacts of neonicotinoids on non-target taxa. A letter signed by 233 concerned scientists (Goulson et al. 2018) called for the urgent further restrictions on their uses, because 'the balance of evidence strongly suggests that the chemicals are harming beneficial insects and contributing to the current massive loss of global biodiversity'.

A rather different form of chemical changes is the tendency for nitrogen enhancement from chemical applications, a theme that has received little attention for insect conservation in Australia but has been shown to be a serious threat to butterflies in Europe. Studies on European butterflies have helped understanding of mechanisms that can cause concern (Warren et al. 2021) in changing either or both of local microclimate or vegetation needed for breeding. Three main mechanisms have been implied: the direct loss of some suitable food plants, wider deterioration of food plant quality, and microclimate cooling. Many habitats frequented by specialist butterflies are naturally nutrient-poor, but nitrogen enrichment led to many plants that require those conditions being replaced by nitrogen-tolerant vegetation. Warren et al. cited the case of the regionally declining European *Lasiommata megera* (the Wall butterfly, Nymphalidae) for which nitrogen deposition is related directly to decline through this modified vegetation growth and the restriction of bare ground on breeding sites (Klop et al. 2015). For some other taxa, vigorous vegetation growth in spring can reduce temperatures within the vegetation and lead to slower growth rates and lower survival of some butterflies that overwinter as eggs or larvae. This microclimate cooling effect, resulting from the combination of warmer temperatures and excess nitrogen, may have substantial impacts on some butterflies (WallisDeVries and van Swaay 2006).

8.3 Alien Species

The enormous number of invasive alien species in Australia, some introduced deliberately (and with increased attention to 'biosecurity' seeking to slow or halt further inadvertent introductions, and those of any species with any known or predicted risk attached) has created novel interactions in most environments, including displacements of native taxa as well as direct physical changes such as loss of natural vegetation and overgrazing by domestic stock. For many native insects, detrimental impacts of invasive alien species as they reach more natural environments have been described largely from investigations on individual species, and wider effects commonly remain more speculative (New 2016). The interactions between alien species and more evident primary threatening processes are also inadequately understood and can depend heavily on local circumstances that render wider generalisations misleading. Invasive grasses can become highly flammable, and routinely burning them can increase flammability of native flora adapted to the more usual and infrequent lower intensity fires, and become more aggressive invaders of natural vegetation as they replace newly flammable insect food plants. Rapid regrowth of some African invasive grasses is promoted by fire, leading to increased invasiveness in many ecosystems (Sands 2018).

Alien insect impacts are varied, and sometimes apparently exaggerated. Fears of non-target effects from predators, parasitoids, pathogens and herbivorous insects introduced deliberately as classical biological control agents for alien arthropod or plant pests have been raised frequently in insect conservation (Howarth 1991, 2001), leading to development of comprehensive 'screening' protocols to assure specificity before release is approved and markedly reducing chances of further losses of native insects from introduced species that essentially become permanent in the receiving environments. Some other insect introductions, accidental or deliberate, have led to wider concerns extending well beyond anthropogenic environments. As elsewhere, some aculeate Hymenoptera are of particular concern in Australia, with feral populations of the European honey bee (*Apis mellifera*) widespread, and domestic use for crop and orchard pollination a mainstay of crop production, largely through migratory beekeeping through which hives are moved to 'track' crops as pollination is needed. A European bumblebee, *Bombus terrestris*, introduced deliberately into New Zealand in the nineteenth century to aid pollination of clover, was first recorded in Tasmania in 1992 (at Battery Point, Hobart, and presumed to have been imported from New Zealand) and has since become widely distributed over the island state. It is one of the most abundant bees in urban areas, and has been claimed to have severe ecological impacts even in remote areas, through competition with smaller native bees and direct destruction of native flowers by chewing through them in order to obtain nectar. Its rapid spread—Stout and Goulson (2000) suggested that 'it is unlikely that its spread will be halted by human interaction'—was accompanied by early expressions of concern for adverse impacts (Hingston and McQuillan 1998). Not all those impacts are obvious; increasing predominance of several so-called 'sleeper weeds', invasive species that are usually rather scarce, has been attributed to increased pollination by *B. terrestris*, and their emergence leads to increased competition with native flora, in addition to the nectar-robbing impacts noted above. Debate over the values of *B. terrestris* for commercial pollination in Tasmania (Parliament of Australia 2017) indicates the complexity of harmonising conservation concerns with human livelihoods and needs, overlapping with continuing debates over the deliberate spread of European honeybees (*Apis mellifera*) in Australia. The latter now has enormous feral populations, and the spread of both species poses serious concerns in affecting the fate of native bees and other pollinators and the native flora on which they depend. The most numerous visitors to apple blossoms in orchards in Tasmania were honeybees (>90% of visits) and *B. terrestris*, with native bees making only 3.3% of visits. Altogether 26 species of native bees were collected in the four orchards surveyed, and only 10 of these were found on apple blossom. The native species depended predominantly on flowering native vegetation and alien flowers in and around the orchards (Prendergast et al. 2021). Such habitat enhancement was a key component of conserving native bees (and by analogy other pollinators) in this and other crop environments based on plants for which native pollinators may play little part in assuring the levels of pollination needed.

Concerns arising from alien bees extend to other nectar-feeders, for example the critically endangered Swift parrot (*Lathamus discolor*) has been noted as possibly affected by *B. terrestris* in Tasmania through competition for eucalyptus nectar. Such concerns include continuing pressures to allow commercial migratory beekeeping (on which the livelihoods of many people depend) to occur in and near national parks and other protected areas, in part compensating apiarists for destruction of previously available nectar supplies as the enveloping landscapes are progressively denuded. Repeated applications to introduce *B. terrestris* to the Australian mainland, predominantly as a pollinator for glasshouse tomatoes, have been opposed strenuously and its listings as a Potentially Threatening Process (Victoria) and a Key Threatening Process (New South Wales) have stimulated surveillance to detect any incidences, with fears including that, should the bee be allowed to be introduced, escapes from greenhouses are almost certain to occur and lead to the bee becoming invasive.

Vespula wasps, as introduced social pests, have significant impacts on native insects in New Zealand (Beggs 2001), and two species have become highly invasive in Tasmania—the European wasp (*Vespula germanica*) from 1959, when it was probably introduced in cargo from New Zealand, and the English wasp (*Vespula vulgaris*) from 1995. Both are generalist predators that feed on a wide variety of invertebrates and also compete with them for resources such as honeydew as a key source of carbohydrates. Severe impacts have been implied for southern Australia: for *V. germanica* in Tasmania, Spradbery and Maywald (1992) reported substantial local reductions in invertebrates when wasp numbers were high. Despite some differences in habitat preferences, both wasps are implicated in declines of the endemic Ptunarra brown butterfly (*Oreixenca ptunarra*, Nymphalidae) in Tasmania. Once wasps were controlled, butterfly numbers rose, and Potter-Craven et al. (2018) recommended that wasp control efforts should be pursued at key butterfly sites across its range in order to lessen chances of extinctions. Additional conservation measures included provision of buffer zones around key grassland areas, and consideration of butterfly translocations to other sites.

Concerns over alien ants augment the general fears of impacts of social Hymenoptera (New 2012) on the biota of areas they colonise. For any such alien species, effective surveillance and early eradication as an ideal counter to their arrival, is a complex and uncertain exercise, and can best be attempted as soon as possible after detection and in the hope that the target is still highly localised and before it has spread (or been spread) into the wider receptor environment. One of the longest and most expensive such campaigns in Australia has been against the Red Imported Fire Ant (RIFA, *Solenopsis invicta*), initially in Queensland. Early efforts (discussed by New 2012) flowed from discovery of this now widespread neotropical ant in the port of Brisbane (February 2001), from which spread led to many infestations in south-east Queensland and development of a National Eradication Plan. However, a recent (4 March 2021) new infestation in Brisbane (of a genetically distinct ant strain) led to finding of the main nest by odour-detecting dogs. Continued surveillance and control efforts are integral to success of such programmes, as demonstrated by eradication of a RIFA infestation in Sydney (found March 2014, declared eradicated on March 2016) and a continuing (at June 2021) campaign against an infestation discovered in November 2019 at Fremantle, Western Australia. *S. invicta* is regarded as one of the most invasive insects to reach Australia, and its eradication strongly in the national interest. Documenting any such instance as fully as possible is a wise investment for future cases.

A successful eradication campaign in New Zealand involved a butterfly that could probably arrive in Australia, and demonstrated the measures needed in undertaking such an exercise. The Palaearctic Large white butterfly, *Pieris brassicae* (Pieridae) was discovered in Nelson (in the north of the South Island of New Zealand) in May 2010. It had been recognised earlier as potentially harmful, and was listed in 2001 as an 'Unwanted Organism' under the New Zealand Biosecurity Act, as likely to invade the country. The major concern was the polyphagous feeding by larvae on Brassicaceae, with most of New Zealand's 83 native species of the family already threatened by habitat loss and other herbivory, with 66 of them accorded a formal threat status. Likelihood of *P. brassicae* spreading was endorsed by the earlier invasion of the closely related Cabbage white (or Small white, *Pieris rapae*) that spread throughout the country in only 5–8 years and has also become widespread in Australia since it was recorded in 1939. At the time it was found in New Zealand, *P. brassicae* occurred over an area of about 100 km^2, mostly in urban residential gardens. The eradication campaign (Phillips et al. 2020) involved considerable public engagement and cooperation in reporting sightings, searching for the butterfly and its early stages, removing host plants, insecticide use against eggs and larvae, searching for pupae, capture of adults and augmenting populations of parasitoid wasps. Extensive cooperative efforts, with continued surveillance and attention led to the last report of *P. brassicae* in the country on 16 December 2014 and official declaration of its national eradication on 22 November 2016, after a four-year dedicated programme.

Not all 'alien' insects are imported to Australia. Concerns may occasionally arise from mixing of native populations or hybridisation from uncritical or accidental translocations or releases—such as escapees from captive stocks, releases of surplus stock raised by collectors, or ceremonial releases at weddings, notably of butterflies outside their natural within-Australia range.

Introduction of pathogens through alien vectors is discussed most frequently for diseases of commercial bees, as a direct threat to the apiary industry, but any such organisms could have potential non-target effects and merit as much concern as the more familiar alien taxa. Known pathogens and vectors can be included in biosecurity operations to attempt to prevent their introductions. Perhaps the most notable and widely publicised recent example in Australia is the Varroa mite (*Varroa destructor*), a devastating pest of European honeybees and leading to colony collapse and severe impacts on the pollination industry. Australia is the only continent with bees that is still free from Varroa, and its arrival in New Zealand (in 2000) strengthens the likelihood that it will indeed reach Australia, most likely through a major port. Early detection and prevention of spread, perhaps aided by monitoring of 'sentinel hives' close to such ports, may aid containment (Hafi et al. 2012).

Spread of alien plants affects particular insects in different ways. In south eastern Australia, spread of Chilean needle grass (*Nassella neesiana*) is a threat to native grasslands as it displaces less competitive native taxa such as *Danthonia*. It is a declared noxious weed, a status that obliges attempts to eradicate it. However, it is also likely to be an important larval food plant for the Golden sun-moth (p. 200), so that controlling *Nassella* could be a threat to this endangered moth (Richter et al. 2013) and its continued presence be critically important for the moth on some sites. In south east Queensland, spread of the ornamental South American vine Dutchman's pipe (*Aristolochia elegans*) into subtropical forests is implicated in decline of the Richmond birdwing butterfly (*Ornithoptera richmondia*, Papilionidae, p. 204). The vine attracts female butterflies to oviposit, but is toxic to the feeding larvae and constitutes an ecological trap. Removal of the alien vine and its replacement by enhancing availability of the predominant native foodplant vine (*Pararistolochia praevenosa*) by plantings of nursery-grown stock is a key facet of conservation management for the butterfly (Sands and New 2013).

More widely, replacement of native mixed eucalypt and other forests with plantations of alien trees, most notably with Monterey pine (*Pinus radiata*) as a predominant softwood crop, is a major contributor to loss of native forest habitats. Subsidiary impacts include pines invading other forests, shading out native low-growing plants and affecting decomposition processes by deposition of pine needles on the forest floor, so affecting communities of epigaeic invertebrates.

8.4 Fire

Variables associated with deliberate anthropogenic burns include area, frequency, severity and intensity, all superimposed on any more natural sporadic fire regimes (New et al. 2010; New 2014). Fire is a natural component of the Australian environment, and a process essential in recycling and regeneration in many ecosystems, and to which much of the biota has become adapted to reflect the highly patchy variations in fire incidence and intensity that have essentially helped to sustain environmental heterogeneity over much of the continent. Major concerns as threats to native insects are largely from the influences of increased fire severity flowing from climate changes and of increased 'control burning' of far larger areas, with high intensity hot burns and at times of the year that increase the vulnerability of taxa present, with both involving direct destruction of critical resources including loss of natural refuges. As with other putative threats, the impacts of fire, whether natural or imposed, on native insects vary greatly. The extent and impacts of bushfires in Australia are exemplified by data on the 1919–1920 'megafires' in New South Wales, for which Hyman et al. (2020) reported that the fires burned 7% of the state, including 2.7 million hectares of national parks—with 57 national parks having >99% of their area affected. Key ecosystems affected were rainforest (35% of the state's rainforest area lost), wet sclerophyll forest (41%) and heathlands (53%). Many insects and other invertebrates are likely to have suffered accordingly. Examining validated specimen records of several invertebrate groups in the extensive collections of the Australian Museum, Sydney, Hyman et al. found that all known occurrences of 29 species were within the fire areas, and a further 46 species had at least half their known occurrences within those areas. In general, vulnerability of insects to fire may be associated with (1) low mobility, such as flightlessness (such as some dung beetles); (2) size, whereby large-bodied species may be less susceptible through ability to attain refuges; (3) especially flammable substrates such as leaf litter; (4) unusually intense fire affecting biotopes that are not normally burned by lower intensity 'natural' fires: damp forest gullies harbouring forest drosophilid flies were discussed by Hyman et al., and (5) narrow range endemics, especially in restricted flammable biotopes. The primary impact is simply rapid death by cremation, or heat as an 'instant' reduction in population size, but far longer-term indirect impacts result from habitat loss and change, and cascade effects in changed communities. Numerous local extirpations and lowered survival rates are likely to occur. However, in Hyman et al.'s (2020) words, referring to the above criteria for vulnerability 'Thousands of species of invertebrates meet these criteria, many of them likely to be undescribed, and the risk of a wave of extinctions is very real'. Further survey is necessary for most species they investigated, simply because their recorded distributions may not completely represent their full range as a foundation for assessing changes. Nevertheless, in this example those species with their entire reported range overrun by the bushfire merit priority for assessment of their post-fire status.

Much of the range of the Alpine silver xenica butterfly (p. 129) on Mt. Buffalo, Victoria, was severely burned by severe wildfire that enveloped much of the plateau in 2003 but the butterfly has persisted and appears not to have contracted much in range (Dunn 2019). An out-of-control management fire burned a key swamp area inhabited by the Ancient greenling damselfly (p. 197) on Wilsons Promontory in the late autumn of 1987. Small numbers of the damselfly survived, apparently due to a fire prevention management action that involved mowing a roadside firebreak that fortuitously included part of the edge of that swamp. This edge of moist green vegetation apparently lowered intensity of the fire in that area, and enabled larvae to survive in the deeper wet substrate, and later develop to increase the population slowly over the next few years (New 1993a). That timing was entirely fortuitous, but in planning management burns to restore or regenerate habitats for specific insects, knowledge of the species' biology becomes critical to optimise timing and chances for the desirable outcome. In the mid-1990s, the small isolated urban sites harbouring the Eltham copper butterfly (p. 202) near Melbourne were gradually undergoing succession and weed invasions and rendering them progressively less suitable for the butterfly, leading to a consensus management decision that burning might be the best tool to regenerate the areas by reducing alien weeds, promoting growth of the sole larval food plant (*Bursaria spinosa*) and 'opening' the tree canopy. To protect the butterfly, only a narrow seasonal window for burning was available, namely in late summer when part-grown caterpillars were close to stopping feeding and moving underground to overwinter in nests of their host ant, *Notoncus* sp. (New et al. 2000; New 2011). Loss of all food plants by burning would not then affect the larvae greatly as they feed little during the autumn and winter seasons, and the *Bursaria* would be expected to regenerate over that period to provide fresh food by spring. Further, the burn had to be in daytime, to avoid cremating the nocturnal larvae when they emerged from the ant nests to ascend the plants and feed. Additional precautions were needed to control fire intensity on those small sites surrounded by housing, with weather conditions critical to avoid fire spread, and the few small 'hotspots' of larval abundance dampened down before the fire was lit. This exercise was successful, entirely due to the detailed planning based on sound knowledge of the species' biology—in conjunction with the dedication and experience of the fire teams and community support. Small areas of key habitats, especially when they harbour insects with limited dispersal powers or opportunity to move elsewhere may need careful protection from fires, or for deliberate burns to be planned very carefully.

For isolated patches in, for example, agricultural or urban landscapes, accidental burns are a serious concern. Across agricultural landscapes in the Western Cape, South Africa, isolated remnant fragments of native renosterveld are important butterfly habitats (Topp et al. 2021) and prescribed burns at intervals help to regenerate the plant diversity needed to foster butterfly diversity, but sensitive management with both fire and livestock grazing is necessary to sustain landscape-level diversity. In that example, partial burning of areas about every decade might benefit low-mobility butterflies by creating gaps (by shrub removal) and increasing plant diversity.

The intricate interactions between fire and insects are illustrated well by the needs of Australia's largest native bee, the Golden carpenter bee (*Xylocopa aerata*, Apidae). The bee is now believed to be extinct on the mainland of South Australia (where it was recorded last in 1896) and Victoria (with the last record in 1938, when its entire known range in the Grampians National Park was devastated by fire), although a few scattered populations persist in New South Wales. Its main stronghold is now on Kangaroo Island, South Australia, where recent surveys (Glatz et al. 2015) indicate that its distribution has contracted toward the west of the island. The solitary bee nests by excavating tunnels in the soft pithy wood of dead *Banksia* trunks and larger branches and the flower spikes of grasstrees (*Xanthorrhoea*). Supply and suitability of both these substrates is related to fire and each fills rather different needs for the bee population. Fire destroys both. After fire, *Xanthorrhoea* can undergo mass flowering, providing abundant flower stems that become suitable for the bee to exploit around 2–6 years post-fire, after which they decay. *Banksia*, conversely, takes far longer—15 years or so to grow, and often far longer to die and become suitable for bee nesting, with the aid of 'white rot' fungus softening the dead wood. Younger or living trees are not usable. From their surveys, Glatz et al. found that extinctions of the bee from burning could flow from three distinct, and sometimes interlinked, causes: (1) complete burning of isolated habitat patches that cannot be recolonised; (2) frequent burns that do not allow sufficient time—at least 20 years—for regrowth and death of *Banksia*; and (3) burns of large areas of continuous vegetation that remove *Banksia* but also synchronise availability of *Banksia* and *Xanthorrhoea* stalks on different time scales. Recommendations for management thus included (1) refraining from prescribed burns in areas where the bee is present; (2) no, or very infrequent (intervals of more than 20 years) localised burning in priority areas with suitable *Banksia* but no bees; together with (3) winter relocations of active *Xylocopa* nests from areas targeted for prescribed burns despite the presence of the bee. Much of the above information reflected study undertaken after much of the bee's habitat was destroyed by a severe fire in 2007. Deployment of artificial nest sites, made of balsa wood to forms resembling *Xanthorrhoea* stalks and *Banksia* trunks to augment this critical resource, was successful, and both types were readily colonised and used by the bees (Hogendoorn et al. 2021). However, almost all (an estimated 95%, including all artificial nesting substrates in all areas that had been successfully supplemented) of *Xylocopa* habitat on Kangaroo Island was destroyed in the massive fires of 2019–2020, so nullifying an important, innovative and very promising conservation campaign (Hogendoorn et al. 2021). Subsequent searches revealed nests in several unburned sites, and further use of artificial nests is viewed as the foundation of longer-term conservation effort for the bee: Hogendoorn et al. noted that natural nesting substrates in *Banksia* trunks will not now be available until around 2055, so that unburned areas must be retained for at least 35 years, and the bee further 'buffered' by not allowing commercial apiary into those areas where honeybees would compete for very limited nectar supplies.

A key principle of using deliberate fires in insect conservation management is to employ it as 'mosaic' burning, whereby habitat heterogeneity and patchiness can be sustained to the greatest possible extent (New et al. 2010) and as a process that—at least to some extent—minimises the more severe outcomes of many wildfires and in part mimics the regimes of 'cultural burning' long practiced by the First Peoples. It may be necessary to protect small or isolated critical areas entirely from burning, and to ensure that adequate buffer zones are retained to reduce accidental spread of fire. Where this is not possible, alternative management may need consideration, with perhaps increased funding and practical help needed, and provision of suitable refuges also considered where possible. The principle of 'mosaic burning' may in some contexts become 'micro-mosaic burning' where only very small areas are burned across an area. In extreme cases it may be possible to only burn individual food plants, as done to rejuvenate individual tussocks of the sedge *Gahnia filum* for larvae of the Yellowish skipper butterfly (*Hesperilla flavescens flavescens*, Hesperiidae) near Melbourne (Relf and New 2009).

8.5 Climate Change

Climate change can be viewed as an additional stressor, superimposed on the more tangible changes evident and projected from the major components of habitat threats such as landscape alterations and impacts of invasive species (Lindenmayer et al. 2010). With great uncertainties over these compounded effects, climate change must be addressed together with other key environmental problems rather than as an isolated phenomenon, and Lindenmayer et al. listed several key strategies to enhance biodiversity conservation in Australia in the face of rapid climate change. These involved (1) preparing for the impacts of major natural disasters; (2) significantly increasing off-reserve conservation efforts, including fostering connectivity; and (3) enhancing the existing reserve network, all of which need attention at multiple scales, together with adaptive monitoring and management. Climate change is regarded widely as one of the most pervasive and unavoidable influences faced by natural environments, and also one of the most difficult and uncertain themes for which to project impacts in other than rather general terms as key habitats are affected and even lost, together with their associated insects. Thus, Janzen and Hallwachs (2016, 2021) referred to the losses of insects from the tops of tropical mountains in Costa Rica because the cloud forests that had long protected those environments were burned off by increasing dryness and heat from climate change—noting that those insects may variously become extinct, disappear locally but persist on other mountain tops, or persist in changed densities and relationships, and some may adapt to the changes that occur. They also commented that the proportions of species that might be partitioned among any of these options are quite unknown. Analyses of changes in distributions of 66 species of bumble bees across North America and Europe suggested that recent climate changes contributed to widespread declines related to the extent and frequency that historically observed climate limits (especially temperature) were approached or exceeded (Soroye et al. 2020). Calamitous impacts from extreme weather events are expected to become more frequent, with sporadic wildfires likely to become more intensive. The recent

(1919–1020) megafires of unprecedented intensity in southern Australia, for example, may be the forerunner of further such events and their severe outcomes for insects and other invertebrates are enhanced by impacts far from the areas actually burned—such as through ash deposition and runoff into streams and rivers. Saunders et al. (2021) discussed how little is actually known of fire effects on insect communities.

Insect population trends attributed to climate change are often reasonably clear, although not all observations attributed to climate impacts can be separated clearly from concurrent effects of other threatening processes (Forister et al. 2021). Themes relevant to insect conservation, and much more widely applicable, include (1) changed distributions of species in response to warming, with strong likelihood of elevational and latitudinal range changes as insects move upward or poleward to coincide with optimal thermal regimes, and—in many cases—implication that those species may be forced to vacate the most unsuitable parts of their existing range as those become progressively unsuitable, so that range changes become a balance between colonisations and extinctions; (2) changed synchronicities in both season and space with critical resources or between any interacting species that may respond differently to a given temperature change; and (3) local or more widespread changes in community diversity and composition and, hence, in wider interactions flowing from interspecific differences in extents and rates of response to change. In recognising the widespread incidence of poleward shifts in butterfly distributions in the northern hemisphere, Parmesan et al. (1999) also noted that the now highly fragmented landscapes encountered might render this difficult for all but the most efficient colonisers that are capable of traversing inhospitable terrain. Particular concerns have been expressed for insects in contexts such as (1) present high elevation alpine species, where both terrestrial and freshwater species have 'nowhere to go' if they are unable to tolerate warming of their current environments (from which lower elevation insects might escape by moving upward), and undergoing additional pressures by possible competition or other interactions with species invading those areas from below—some 'listed' alpine species may be especially susceptible; and (2) entirely coastal species for which relatively small rises in sea level would inundate the complete current range and resources. One such example is the localised endemic skipper, the Black grass-dart (*Ocybadistes knightorum*, Hesperiidae), for which much of the small range in New South Wales is less than five metres in elevation and some submersible by even current 'king' tides. This endangered butterfly is restricted to these small semi-saline areas, often near mangroves, where its sole larval food plant *Alexfloydia repens*, itself highly localised, can thrive (Andren and Caven 2012, 2014).

The continued nexus between insects and obligatory or preferred host plants is a leading concern, with differential responses to climate changes having potentials to decouple long co-evolved synchrony and specificity, and to create a multitude of novel interactions and opportunities as community compositions of both plants and insect consumers change, and the wider spectrum of players, including predators and parasitoids, is modified. The current overlapping ranges of host plants and their obligatory insect herbivores could shift differently as climate changes occur.

Implications of such asynchronous responses can become complex, and hamper the ability of an insect to 'track' food resources (Pelini et al. 2010). Very commonly, the factors that regulate host plant specificity and the dietary flexibility (if any) of supposedly specialist consumers is not known. As Moir (2015) observed, thousands of insect species are potentially affected and the widespread assumption that simply conserving the host plant will also protect their dependent insect species is often not valid. She suggested that conservation actions are needed urgently, because (1) large numbers of host plant species are threatened imminently by climate change and other factors, and (2) populations of host-specific insects may decline faster than the host plant populations, and plant populations may decline to below the threshold supply needed to sustain a viable consumer population. Management may take many forms and include both *in situ* and *ex situ* components that extend to many aspects of habitat remediation and tracking. Local adaptations of herbivorous insects to different food plants may create complex mosaics of interactions across a species' range and as the dynamics of individual hosts change continually. Local climate impacts on insects such as through urban heat island effects deserve further investigations in Australia, and their alleviation links strongly with provision of open green spaces in urban areas.

Australia is generally rather flat, and lacks a broad range of elevations—the highest point (Mt Kosciuszko, New South Wales) is only 2228 m above sea level. As Hughes (2011) pointed out, opportunities for upward movements of species in response to warming are very limited by this short gradient, and the alternative movement response necessitates long overland range changes. With even a low level of warming many Australian ecosystems are likely to undergo changes, but Hughes considered that especially high impacts could occur on biodiversity of the alpine zone, freshwater and coastal fringe systems (including wetlands), and the biodiversity hotspots of the north Queensland wet tropics and south-western Western Australia (as well as for coral reefs). High elevation streams are experiencing severe abiotic changes, encompassing temperature, flow, water oxygen levels and chemistry and exposure to ultraviolet radiation, providing a considerable variety of stresses to their insect inhabitants (Birrell et al. 2020). For these, and many others, long-term projections include considerable change and loss, but short-term changes such as from sudden temperature extremes, severe droughts, heavy precipitation, flood and fire related to climate changes, can occur very rapidly. Concerns proliferate that many insects exposed to such events are facing circumstances unprecedented in their evolutionary history, and to environments beyond their adaptive limits (Harvey et al. 2020), also involving breakdowns in interactions and trophic networks. Such concerns have been expressed for the insects of all their major habitats, and are exemplified by alpine insect distributions contracted directly by change and from novel interactions from taxa colonising from lower elevations, as noted earlier, and including considerable vegetational changes such as woody vegetation replacing specialised alpine herbaceous and grassland systems. Seasonal changes include earlier spring thaws, declining snow cover, and contraction of bogs from changed hydrology. Changes in drought regimes affect water levels and flows across regions such as the Murray-Darling Basin, and sea level rises affecting tidal levels modify coastal and mangrove vegetation. Many such impacts on insect habitats (with potential threats to endemic insects) remain to be investigated, and no formal monitoring is in place for most of those environments. Geographically restricted areas (such as the alpine region) are refuges for archaic endemic insects, and almost certainly vulnerable to changing climates. Many montane insects may suffer high risks as already highly constrained in distribution and ecological tolerances and perhaps as poor competitors with species colonising their ranges. Topographical variety and physical heterogeneity contribute to maintaining a variety of microclimates and, thus, an equivalent variety of insect life. Writing more generally on threats to Australian fauna, Allek et al. (2018) noted that climate changes and extreme weather events posed greater danger in the east of the country than elsewhere with—in their example—many amphibians very sensitive to temperature change. A similar claim could be valid for the numerous cool-adapted and higher elevation insects of the eastern mainland and Tasmania. Many of these taxa occur only within rather narrow cool temperature ranges likely to contract substantially if warming occurs, and many are also suspected to be intolerant of warmer conditions.

Range changes have been inferred for running water invertebrates in eastern Australia (Chessman 2012). Vulnerability can occur from increased water temperatures (as in alpine taxa, above and reflecting more widely that many aspects of insect biology are temperature-governed), changes to hydrological regimes, constraints on dispersal and changed interactions with non-climate threats. Members of cool-adapted Gondwanan lineages may be especially susceptible, and numerous such aquatic groups are restricted to Tasmania and/or the southeast mainland. Range contraction may be the only, or predominant, response to climate changes for these insects. Predictors of susceptibility have involved both water temperature and flow regimes, and Chessman suggested that (1) rising air temperatures warm streams to the disadvantage of cool-water adapted insects unable to tolerate warmer temperatures and (2) lowered stream flows from changed precipitation, transpiration and evaporation may affect species adapted to living in flowing water. Both traits appeared to be important.

Distributional changes, whether in elevation or latitude, can succeed only if critical resources, predominantly food, are available in the areas to be colonised. For insects, suitable food plants or prey/hosts are the most critical needs, with any management directed to assuring a two-partner association. More complex systems occur. For example, many myrmecophilous lycaenid butterflies must also encounter a specific mutualistic ant in an obligatory tripartite relationship. Four of the seven endemic Australian butterflies identified as at particular risk from climate changes by Beaumont and Hughes (2002) are such Lycaenidae with narrow geographical and climatic ranges. The immense variety of intricate and specific associations between butterfly-plant-ant can render practical conservation difficult (Pierce et al. 2002; New 1993b), but with the universal requirement that all three species involved are needed. For some ants, as demonstrated for the *Myrmica* host species of Large blue butterflies (*Maculinea*) in Europe, ground temperatures as influenced by aspect and vegetation cover may need to be kept within rather narrow ranges.

Early trends of climate change impacts were observed mostly as poleward range shifts of well-documented butterflies in the northern hemisphere, and were paralleled later by similar trends in Odonata in Europe, with implications that such changes are likely be occurring undetected in other insect groups. Simultaneously, records of butterflies and Odonata, especially in Britain and northern Europe, demonstrated earlier appearances in spring, as a seasonal shift. In Sweden (Flenner and Sahlen 2008), strong responses for dragonflies were evident in only a decade, leading to rapid changes in assemblage composition and species abundance in boreal forest lakes. The trends included some rare species becoming more frequent and widespread, and some previously widespread species becoming more selective in their habitat range. Such changes may affect the values of Odonata as a useful indicator group, and need careful consideration of this in conservation planning (Flenner and Sahlen 2008). Predicting future trends has involved 'modelling', often drawing on projected changes of the current climate envelope under various defined levels of change. A species' climate envelope is the climate parameters that circumscribe its current distribution and are thus deemed favourable but, despite its attractiveness, this approach has many shortcomings (Maes et al. 2010). In Belgium, trends for butterflies, dragonflies and grasshoppers were examined under a series of projected climate scenarios and models. Butterfly and grasshopper densities were projected to decrease in all scenarios examined, and locations for greatest species richness projected to move to higher elevations (Maes et al. 2010). Dragonflies differed somewhat, in that species-rich locations moved upward only in the less severe scenarios. Impacts on different insect groups can clearly differ, but an important outcome from Maes et al.'s study was that changes in sites important for different insect taxa will not always coincide, so that the overall area needed for protection in the future must be increased. New protected areas would be needed to include biotopes that are, or are projected to become, suitable for those species in the future and positioned so that affected species can readily move to colonise them. Those reserves must provide for future trends effectively, rather than simply enhancing present-day conservation. However, planning for the variety of scales and locations of changes and, indeed, working out what is really causing observed changes, poses many uncertainties. As Hardy et al. (2014) discussed, changes in an insect's population, distribution and range are 'a great deal more complex than hitherto considered'. That thoughtful essay, based on the butterflies of a region of northwest England, also noted the distinctions between elevational and latitudinal shifts in distribution. In following a resource-based concept of 'habitat' (p. 91), Hardy et al. suggested that new higher elevation sites are more likely to be of marginal quality, supporting smaller populations. The movement may be easier than moving poleward across highly altered landscapes; higher elevation sites may be gained more easily, but are also more easily lost as conditions change.

The difficulties of predicting effects of climate changes on poorly-known species are formidable, and the biases in taxon coverage and geographical representation in studies noted by Andrew et al. (2013) still persist widely. Much of the detailed awareness of climate effects comes from long-term monitoring of pest insects, and wider ecological studies are needed (Larson et al. 2019). The trends evident from

Europe and North America are indeed likely to be paralleled elsewhere, but the great variety of accounts—Andrew et al. noted 1703 publications between 1985 and August 2012—reflected studies across many habitats (in many papers, not stated precisely), taxa, and circumstances, including anthropogenic pressures. And, with the most commonly investigated trends being in distribution, phenology and abundance, accompanying impacts on development, physiology and wider environmental tolerances are easily overlooked. Climatically-related variables, notably temperature, precipitation and photoperiod are all strong influences on insect development, with phenological changes affecting key interactions with resources and other community members. In addition to the gradual but inexorable warming and other climatic changes, extreme weather events—flooding from heavy precipitation, droughts, heat waves and markedly unusual temperature episodes—are also increasing and likely to drive many aspects of insect life by further decoupling interactions and causing direct mortality, sometimes linked with other tendencies such as increased fire frequency and intensity. Although these have not yet been studied extensively in Australia, the example of the Karner blue butterfly (*Lycaeides melissa samuelis*, Lycaenidae) and its obligate host plant, a Blue lupin (*Lupinus perennis*) in North America illustrates well that adaptive management strategies that include considerations of extreme weather events are needed urgently (Patterson et al. 2020), with continued tracking of environmental conditions to form a precautionary management process that is readily adaptable to changing circumstances. The recent history of another ant-associated lycaenid, the Brenton Blue (or Brenton Cupid) in South Africa exemplifies a scenario that is perhaps not unlikely for some threatened Australian insects. *Orachrysops niobe* has been known recently from a single small dedicated reserve in the Western Cape Province, where it is the focus of a strong conservation programme led by a committee including all relevant stakeholder groups. A prolonged drought (2014–2017) was associated with a 93% decline in adult numbers seen, and a severe fire in June 2017 burned much of the reserve. The butterfly has not been seen since the fire and, although categorised as Critically Endangered, it is not yet clear whether it still exists (Edge 2021). Monitoring is being continued.

Mechanisms of insect declines attributed directly to climate change have rarely been elucidated in detail. Forest floor beetles in New Hampshire, United States, were sampled by window traps in 1973–1977 and again in 2015–2017 (Harris et al. 2019). Over that period numbers declined by >80%, and 15 families captured in the first series were not rediscovered. Greatest declines were among predatory taxa. The major drivers of decline were reduced duration and depth of snow cover. Decreased depth of snow led to soil freezing to a greater depth, so that winter warming equated to harsher conditions for the ground fauna. Whether the declines observed were driven primarily by lower survival (especially of small-bodied taxa) in these conditions, or more indirectly (such as by reduced prey for predators) was uncertain—but shorter warmer winters could also increase activity of winter-active predators. No other factors involved in habitat loss or degradation were implicated in the declines found, and Harris et al. concluded that a temperature increase of only 1 °C can have major impacts on forest insect diversity and hence community structure and ecosystem processes. A central dilemma in interpreting impacts of climate change on many insect communities relates to either (1) the positive responses of increasing temperature and growing season, perhaps increasing the number of generations possible each year, enhancing reproduction and reducing periods of dormancy and (2) the negative impacts of changing precipitation that decrease winter snow cover and combine with higher temperature effects to decrease survival.

One possible practical counter to climate change (or, indeed, to wider habitat loss) is some form of 'assisted migration' of especially significant species to areas that are predicted to become or remain suitable as conditions change. Thus, attempts were made, unsuccessfully, to re-introduce the Brenton Blue to a formerly occupied site from where it disappeared in the 1970s.

Translocations of insects are advocated widely in insect conservation, for a variety of purposes—such as (1) increasing the number of populations as movements to new (unoccupied) sites or as re-introductions to sites where the species has been extirpated and that have now been protected and/or restored; (2) augmenting small populations, by adding individuals to already occupied sites to increase population viability; and (3) 'rescue' by salvaging individuals from sites that are to be destroyed, and using them to found captive breeding populations or for direct release elsewhere, as above. In principle, either field-collected or captive-reared specimens can be used in translocations: for the former, care is needed not to increase vulnerability of the donor population by removal of individuals. As contrasting examples: (1) reintroduction of the Lord Howe Island stick insect (p. 200) to Lord Howe and its near neighbours is based on captive-reared material, whilst (2) possible future movements of Green carpenter bee (p. 204) to the Grampians National Park in Victoria could depend on translocations of occupied artificial nests. For the bee, modelllng of climate changes (Santos et al. 2020) implied that the Grampians would indeed remain suitable to support the bee and the several key plant species needed for nesting, and chances of success are likely to be high (Hogendoorn et al. 2021).

Within Australia, most insect conservation currently focuses on in situ measures, emphasising occupied site selection, protection, and restoration and other management as needed. Translocations to novel sites are rare, but have been contemplated for several notable species that are known to survive on small areas. The Golden sun-moth is one such species (p. 204), and the successful case of emergency 'salvage' transfer for the Bathurst copper butterfly (*Paralucia spinifera*) (Mjadwesch and Nally 2008) demonstrated the wisdom of including such considerations in planning conservation of individual species (New 2011). The Bathurst copper case emphasised the importance of both background ecological knowledge to inform the process and the cooperation between all the parties involved, with the common purpose of success: the butterfly persisted for at least a decade after translocation (Nally and Mjadwesch 2019). However, unlike the conservation programmes for weta in New Zealand, translocations are not yet a cornerstone of most recovery plans for Australian insects. A major threat to weta is predation by introduced mammals, so a key to their conservation is access to predator-free environments, including small islands from which predators have been eradicated. Captive-bred weta have been introduced successfully to such islands within the historical range of each species (Sherley 1998; Watts et al. 2012), and a major conservation need is now to prevent re-invasion by stoats, rats and others.

8.5.1 Habitat Offsets

A related theme in considerations of assisted migration of insects is the principle of 'habitat offsets', in some way compensating for loss of natural, habitat to development by allocation of alternative suitable area for the focal species. Brought forcibly to focus in Australia from the plight of the Golden sun-moth (*Synemon plana*, p. 200), this step may be viewed by developers as a short-term avenue toward justifying land clearing by undertaking such compensatory activities. Although a definition of 'environmental offsets' (as 'actions taken outside a development site that compensate for the impacts of that development—including direct, indirect or consequent impacts': Australian Government 2007) has for long been in place, further operating conditions have been advocated. Gibbons and Lindenmayer (2007) suggested that offsets might be feasible (1) only for relatively simple vegetation types that can indeed be restored; (2) where the offset must be in place before the original site is lost; (3) where management of the offset site is adaptive; and (4) that the offset site must be secure for the future. Defining the values needed in an offset site for a rare insect can become difficult, but the underlying principle of ensuring 'no net loss' dictates this need. The detailed needs of any focal insect species are generally insufficiently known, so that only a basic framework of the more obvious requirements can be considered. Availability of offsets can be attractive to developers as a cost-effective means of complying with environmental and planning laws, reducing processing times for applications and giving greater certainty over planning requests (BlakeDawson 2008). However, many practical uncertainties persist and the offset process for insects cannot routinely replace loss of primary habitats: the loss of any such areas to development is a serious step that needs careful and individual consideration (New 2009).

8.6 Overexploitation

Whilst the most usual context of concern of 'overexploiting' insects is the increased vulnerability to already rare species by excessive or commercial collecting, most commonly of rare butterflies or beetles, for commercial sale—with desirability increased by formal prohibition of capture—most of the expressed fears for Australian insects have not borne critical scrutiny (Sands and New 2002), especially in relation to the more severe threats that have led to any such accusations.

Other contexts for concern arise occasionally. The take of large numbers of dragonfly larvae ('mudeyes') as bait for recreational angling may occasionally need to be monitored if the localised habitats of notable species (such as *Petalura*, p. 197) are affected. Recent increased interest in 'bush tucker' is associated with increased tourism to inland Australia and acknowledgement and involvement of Australia's earlier custodians. Insects comprised important food elements and although some (such as sugarbag bees, *Tetragonula carbonaria* and related species) can be domesticated for production, others must be harvested from the wild. Honeypot ants (species of *Melophorus* and *Camponotus*) and, especially, 'witchetty grubs' (the large larvae of cerambycid beetles and cossid moths) might be subject to local overcollection in places if market demands increase.

The 'Butterfly House Industry' is founded largely on continuing supplies of pupae, as a 'cash crop', from wild-sourced populations. Although Australian operations are founded on (and restricted to) native species, so avoiding the complications and risks of the international trade widespread elsewhere (Boppre and Vane-Wright 2012), there is currently no suggestion of over-collecting for this purpose.

References

Allek A, Assis AS, Eiras N, Amaral TP, Williams B et al (2018) The threats endangering Australia's at-risk fauna. Biol Conserv 222:172–179

Andren M, Caven MA (2012) The distribution of the endangered black grass-dart butterfly, *Ocybadistes knightorum* (Lepidoptera: Hesperiidae). Aust Zool 36:159–168

Andren M, Caven MA (2014) The conservation status of the black grass-dart butterfly, *Ocybadistes knightorum*: a species at risk from climate change. Aust Zool 37:76–84

Andrew NR, Hi SJ, Binns M, Bahar MH, Ridley EV et al (2013) Assessing insect responses to climate change: what are we testing for? Where should we be heading? PeerJ 1:e11. https://doi.org/10.7717/peerj.11

APVMA (Australian Pesticides and Veterinary Medicines Authority) (2014) Overview report. Neonicotinoids and the health of honey bees in Australia, Canberra

Australian Government (2007) Use of environmental offsets under the Environment Protection and Biodiversity Conservation Act 1999. Discussion paper. Department of the Environment and Water Resources, Canberra

Beaumont LJ, Hughes L (2002) Potential changes in the distributions of latitudinally restricted Australian butterfly species in response to climate change. Glob Change Biol 8:954–971

Beggs J (2001) The ecological consequences of social wasps (*Vespula* spp.) invading an ecosystem that has an abundant carbohydrate resource. Biol Conserv 99:17–28

Birrell JH, Shah AA, Hotaling S, Giersch JJ, Williamson CE et al (2020) Insects in high-elevation streams: life in extreme environments imperiled by climate change. Glob Change Biol 26:6667–6684

BlakeDawson (Nelson R, Day N, Jamieson R, Hill T) (2008) Environmental matters (national) (5/06/2008) Biodiversity: can you bank on it? www.BlakeDawson.com/templates/pubications/x_article_content_page.aspx?id

Boppre M, Vane-Wright RI (2012) The butterfly house industry: conservation risks and education opportunities. Conserv Soc 10:285–303

Braby MF, Williams MR, Douglas F, Beardsell C, Crosby DF (2021) Changes in a peri-urban butterfly assemblage over 80 years near Melbourne, Australia. Austral Entomol 60:27–51

Canzano A, Whitfield J (2008) Discovery of a further population of the Eltham copper butterfly *Paralucia pyrodiscus lucida* Crosby (Lepidoptera: Lycaenidae) in Bendigo, Victoria. Vict Nat 125:178–180

Chessman BC (2012) Biological traits predict shifts in geographical ranges of freshwater invertebrates during climate warming and drying. J Biogeogr 39:957–969

Dunn KL (2019) *Oreixenica latialis* (Lepidoptera: Nymphalidae: Satyrinae) on Mount Buffalo. Victoria: field notes on its local distribution and adult behaviour to assist future conservation work. Calomela 724: 1–24

Edge DA (2021) *Orachrysops niobe* (Trimen, 1862). In: Mecenero S, Edge DA, Staudte HS, Coetzer BH, Coetzer AJ et al Outcomes of the Southern African Lepidoptera conservation assessment (SALCA). Metamorphosis, vol 31, pp 113–114

Flenner I, Sahlen G (2008) Dragonfly community re-organisation in boreal forest lakes: rapid species turnover driven by climate change. Insect Conserv Divers 1:169–179

Forister ML, Halsch CA, Nice CC, Fordyce JA, Dilts TE et al (2021) Fewer butterflies seen by community scientists across the warming and drying landscapes of the American west. Science 371:1042–1045

Gibbons P, Lindenmayer DB (2007) Offsets for land clearing: no net loss or the tail wagging the dog? Ecol Manag Restor 8:26–32

Glatz RV, Leijs R, Hogendoorn K (2015) Biology, distribution and conservation of Green carpenter bee (*Xylocopa aeratus*: Apidae) on Kangaroo Island, South Australia. Technical report, Department of Environment, South Australia

Goulson D (2019) The insect apocalypse, and why it matters. Curr Biol 29:R967–R971

Goulson D et al (2018) Call to restrict neonicotinoids. Science 360:973

Hafi A, Millist N, Morey K, Caley P, Buetre B (2012) A benefit-cost framework for responding to an incursion of Varroa destructor. ABARES research report 12.5, prepared for National Biosecurity Committee, Commonwealth of Australia, Canberra

Hardy PB, Sparks TH, Dennis RLH (2014) The impact of climatic change on butterfly geography: does climatic change produce coincident trends in populations, distributions and ranges? Biodivers Conserv 23:855–876

Harris JE, Rodenhouse NL, Holmes RT (2019) Decline in beetle abundance and diversity in an intact temperate forest linked to climate warming. Biol Conserv 240:108219

Harvey JA, Heinen R, Armbrecht I, Basset Y, Baxter-Gilbert JH et al (2020) International scientists formulate a roadmap for insect conservation and recovery. Nat Ecol Evol 4:174–176

Hingston A, McQuillan PB (1998) Does the recently introduced bumblebee *Bombus terrestris* (Apidae) threaten Australian ecosystems? Aust J Ecol 23:539–549

Hogendoorn K, Glatz RV, Leijs R (2021) Conservation management of the green carpenter bee *Xylocopa aerata* (Hymenoptera: Apidae) through provision of artificial nesting substrate. Austral Entomol 60:82–88

Howarth FG (1991) Environmental impacts of classical biological control. Annu Rev Entomol 36: 485–509

Howarth FG (2001) Environmental issues concerning the introductions of non-indigenous biological control agents. In: Lockwood JA, Howarth FG, Purcell MF (eds) Balancing nature: assessing the impacts of importing non-native biological control agents (an international perspective). Entomological Society of America, Lanham, MA, pp 70–99

Hughes L (2011) Climate change and Australia: key vulnerable regions. Reg Environ Change 11: S189–S195

Hyman IT, Ahyong ST, Kohler F, McEvey SF, Milledge G, Reid CAM, Rowley JJL (2020) Impacts of the 2019-2020 bushfires on New South Wales biodiversity: a rapid assessment of distribution data for selected invertebrate taxa. Tech Rep Aust Mus 32:1–17

IUCN (2017) Threats Classification Scheme (Version 3.2). http://www.iucnredlist.org/technical-documents/classification-schemes/threats-classification-scheme

Janzen DH, Hallwachs W (2016) DNA barcoding the Lepidoptera inventory of a large complex tropical conserved wildland, Area de Conservacion Guanacaste, northwestern Costa Rica. Genome 59:641–660

Janzen DH, Hallwachs W (2021) To us insectometers, it is clear that insect decline in our Costa Rican tropics is real, so let's be kind to the survivors. Proc Natl Acad Sci 118:e2002546117

Kearney SG, Carwardine J, Reside AE, Fisher DO, Marin M et al (2019) The threats to Australia's imperilled species and implications for a national conservation response. Pac Conserv Biol 25: 231–244

Klop E, Omon B, WallisDeVries MF (2015) Impact of nitrogen deposition on larval habitats: the case of the Wall Brown butterfly *Lasiommata megera*. J Insect Conserv 19:393–402

Larson EL, Tinghitella RM, Taylor SA (2019) Insect hybridization and climate change. Front Ecol Evol 7:348. https://doi.org/10.3389/fevo.2019.00348

Lindenmayer DB, Steffen W, Burbidge AA, Hughes L, Kitching RL et al (2010) Conservation strategies in response to rapid climate change: Australia as a case study. Biol Conserv 143: 1587–1593

Maes D, Titeux N, Hortal J, Anselin A, Decleer K et al (2010) Predicted insect diversity declines under climate change in an already impoverished region. J Insect Conserv 14:483–498

Marks EN (1969) The invertebrates. In: Webb LJ, Whitelock D, Le Gay Brereton J (eds) The last of lands. Conservation in Australia. Jacaranda Press, Milton, pp 102–114

Mjadwesch R, Nally S (2008) Emergency relocation of a purple copper colony during roadworks: successes and lessons learned. Ecol Manage Restor 92:100–109

Moir M (2015) Conservation of Australia's plant-dwelling invertebrates in a changing climate. In: Palutikof JP, Boulter SL, Barnett J, Rissik D (eds) Applied studies in climate adaptation. Wiley, Chichester, pp 107–116

Nally S, Mjadwesch R (2019) Butterfly population persists 10 years after emergency habitat restoration and translocation - UPDATE to EMR feature. https://site.emrprojectsummaries.org/2019/09/30/butterfly-population. Accessed March 2021

New TR (1993a) *Hemiphlebia mirabilis* Selys: recovery from habitat destruction at Wilsons Promontary, Victoria, Australia, and implications for conservation management. Odonatologica 22:495–502

New TR (1993b) Introduction to the biology and conservation of Lycaenidae. In: New TR (ed) Conservation biology of Lycaenidae (butterflies). IUCN, Gland, pp 1–21

New TR (2009) Habitat offsets for insect species conservation: practicality or placebo? J Insect Conserv 13:139–141

New TR (2011) Butterfly conservation in South-Eastern Australia: progress and prospects. Springer, Dordrecht

New TR (2012) Hymenoptera and conservation. Wiley-Blackwell, Chichester

New TR (2014) Insects, fire and conservation. Springer, Cham

New TR (2015) Insect conservation and urban environments. Springer, Cham

New TR (2016) Alien species and insect conservation. Springer, Cham

New TR (2018) Promoting and developing insect conservation in Australia's urban environments. Austral Entomol 57:182–193

New TR (2021) Insect conservation and Australia's inland waters. Springer, Cham

New TR, Yen AL (2012) Insect conservation in Australia. In: New TR (ed) Insect conservation: past, present and prospects. Springer, Dordrecht, pp 193–212

New TR, Van Praagh BD, Yen AL (2000) Fire and the management of habitat quality in an Australan lycaenid butterfly, *Paralucia pyrodiscus lucida* Crosby, the Eltham copper. Metamorphosis 11:154–163

New TR, Yen AL, Sands DPA, Greenslade P, Neville PJ, York A, Collett NG (2010) Planned fires and invertebrate conservation in south East Australia. J Insect Conserv 14:567–574

Parliament of Australia (2017) Risks and opportunities associated with the use of the bumblebee population in Tasmania for commercial pollination purposes. Report from the Senate Environment and Communications References Committee. Commonwealth of Australia, Canberra

Parmesan C, Ryrholm N, Stefanesku C, Hill JK, Thomas CD et al (1999) Poleward shifts in geographical ranges of butterfly species associated with regional warming. Nature 399:579–583

Patterson TA, Grundel R, Dzurisin JDK, Knutson RL, Hellmann JJ (2020) Evidence of an extreme weather-induced phenological mismatch and a local extirpation of the endangered Karner blue butterfly. Conserv Sci Pract 2:e147. https://doi.org/10.1111/csp2.147

Pelini SL, Keppel JA, Kelley AE, Hellmann JJ (2010) Adaptation to host plants may prevent rapid insect responses to climate change. Glob Change Biol 16:2923–2929

Phillips CB, Brown K, Green C, Croft R, Walker G, Broome K (2020) Eradicating the large white butterfly from New Zealand eliminates a threat to endemic Brassicaceae. PLoS One 15(8): e0236791. https://doi.org/10.1371/journal.pone.0236791

Pierce NE, Braby MF, Heath A, Lohman DJ, Mathew J, Rand DB, Travassos MA (2002) The ecology and evolution of ant association in the Lycaenidae. Annu Rev Entomol 47:733–771

Pisa LW, Amaral-Rogers V, Belzunces LP, Bonmatin JM, Downs CA et al (2015) Effects of neonicotinoids and fipronil on non-target invertebrates. Environ Sci Pollut Res 22:68–102

Potter-Craven J, Kirkpatrick JB, McQuillan PB, Bell P (2018) The effects of introduced vespid wasps (*Vespula germanica* and *V. vulgaris*) on threatened native butterfly (*Oreixenica ptunarra*) populations in Tasmania. J Insect Conserv 22:521–532

Prendergast KS, Leclercq N, Vereecken NJ (2021) Honey bees (Hymenoptera: Apidae) outnumber native bees in Tasmanian apple orchards: perspectives for balancing crop production and native bee conservation. Austral Entomol 60:422–435

Relf MC, New TR (2009) Conservation needs of the Altona skipper butterfly, *Hesperilla flavescens flavescens* Waterhouse (Lepidoptera: Hesperiidae), near Melbourne, Victoria. J Insect Conserv 13:143–149

Richter A, Osborne W, Hnatiuk S, Rowell A (2013) Moths in fragments: insights into the biology and ecology of the Australian endangered golden sun moth *Synemon plana* (Lepidoptera: Castniidae) in natural temperate and exotic grassland remnants. J Insect Conserv 17:1093–1104

Sands DPA (2018) Important issues facing insect conservation in Australia: now and into the future. Aust Entomol 57:150–172

Sands DPA, New TR (2002) The action plan for Australian butterflies. Environment Australia, Canberra

Sands DPA, New TR (2013) Conservation of the Richmond birdwing butterfly in Australia. Springer, Dordrecht

Santos AA, Leijs R, Picanco MC, Glatz RV, Hogendoorn K (2020) Modelling the climate suitability of green carpenter bee (*Xylocopa aerata*) and its nesting hosts under current and future scenarios to guide conservation efforts. Austral Ecol 45:271–282

Saunders ME, Barton PS, Bickerstaff JRM, Frost L, Latty T et al (2021) Limited understanding of bushfire impacts on Australian invertebrates. Insect Conserv Divers 14:285–293

Sgolastra F, Medrzychi P, Bortolotti L, Maini S, Porrini C et al (2020) Bees and pesticide regulation: lessons from the neonicotinoid experience. Biol Conserv 241:108356. https://doi.org/10.1016/j.biocon.2019.108356

Sherley GH (1998) Threatened weta recovery plan. Department of Conservation, Wellington

Soroye P, Newbold T, Kerr J (2020) Climate change contributes to widespread declines among bumble bees across continents. Science 367:685–688

Spradbery JP, Maywald GF (1992) The distribution of the European or German wasp, *Vespula germanica* (F) (Hymenoptera: Vespidae), in Australia – past, present and future. Aust J Zool 40: 495–510

Stout JC, Goulson D (2000) Bumble bees in Tasmania: their distribution and potential impact on Australian flora and fauna. Bee World 81:80–86

Topp EN, Tscharntke T, Loos J (2021) Fire and landscape context shape plant and butterfly diversity in a South African shrubland. Divers Distr 2021. https://doi.org/10.1111/ddi.13257

TSSC (Threatened Species Scientific Committee) (2016) Conservation advice *Paralucia spinifera* purple copper butterfly. Threatened Species Scientific Committee, Canberra

WallisDeVries MF, van Swaay CAM (2006) Global warming and excess nitrogen may induce butterfly decline by microclimatic cooling. Glob Change Biol 12:1620–1626

Ward MS, Simmonds JS, Reside AE, Watson JEM, Rhodes JR et al (2019) Lots of loss with little scrutiny: the attrition of habitat critical for threatened species in Australia. Conserv Sci Pract 2019(1):e117

Ward M, Carwardine J, Yong CJ, Watson JEM, Silcock J et al (2021) A national-scale dataset for threats impacting Australia's imperiled flora and fauna. Ecol Evol 00:1–13. https://doi.org/10.1002/ece3.7920

Warren MS, Maes D, van Swaay CAM, Goffart P, Van Dyck H et al (2021) The decline of butterflies in Europe: problems, significance, and possible solutions. Proc Natl Acad Sci 118(2): e2002551117

Watts C, Stringer I, Gibbs G (2012) Insect conservation in New Zealand: an historical perspective. In: New TR (ed) Insect conservation: past, present and prospects. Springer, Dordrecht, pp 213–243

Chapter 9
Facilitating Conservation Progress

9.1 Introduction

Insect conservation in Australia has become acknowledged widely as important, together with awareness that much future progress must occur without precise documentation of either any focal species or of the richness and peculiarities of apparently unusual communities—but also with the realisation that insect life is diverse, largely endemic to Australia (and often to small regions of the continent), threatened, and irreplaceable. Many gaps in knowledge will persist, and conservation priorities must be determined and pursued within that perspective and without waiting for wholly definitive information and detail to be accumulated. Losses of sensitive environments and their endemic insect residents are sure to occur far more rapidly than requisite scientific advances, and an informed precautionary approach to their conservation is necessary to reduce apparent threats and protect those environments.

The numbers of species in most larger insect orders in Australia will remain uncertain, and it is inevitable that concepts of species delimitation will differ amongst specialists, perhaps especially as to the practical implications of haplotype variety revealed by DNA barcode analyses, with likelihood that numerous 'cryptic species' may lurk unidentified amongst those taxa that have been recognised from morphological features alone. Sufficient is known to recognise many of the most unusual and notable insect lineages, species and radiations as foci for further attention, and also to emphasise the relevance of improving knowledge of these as important tools in promoting the worth of insect conservation throughout the Australian populace and incorporating insects also into wider conservation agendas to safeguard Australia's unique biodiversity. Although insects are indeed acknowledged formally in Australian environmental legislations, their conservation needs as perceived by many people are secondary to those of the relatively high profile and charismatic vertebrates, and to vascular plants in popularity. As observed many times, insects are generally treated simply as part of the overlooked biodiversity, so

© The Author(s), under exclusive license to Springer Nature Switzerland AG 2022
T. R. New, *Insect Diversity, Declines and Conservation in Australia*, Fascinating Life Sciences, https://doi.org/10.1007/978-3-030-90134-9_9

often disregarded and considered (if at all!) insignificant. In social environments in which adequate conservation of even the most iconic marsupials, monotremes or birds remains uncertain (with lack of consensus on, for example, conservation needs of the Koala (*Phascolarctos cinereus*) in different parts of Australia, and continuing decline of the Platypus in Victoria), the low profile of most insects seems likely to continue.

That few people heed the conservation of unusual insect species is both a hindrance (through failures to gain wider impetus through wide interest and public advocacy) and a benefit (in avoiding controversy amongst interested parties and providing a united case for conservation need). Almost all cases in which insect species have indeed progressed beyond 'listing' to more practical evaluation and conservation action exemplify this unity, and promotion by well-informed and concerned observers whose continued commitment may be instrumental in any successful outcome. However, the more common fate of most listed invertebrates is simply welcome but transient notoriety through gaining that status and priority, but little further action. For example, Victoria's Flora and Fauna Guarantee Act obliges preparation of an 'Action Statement' for each listed taxon, with revisions at prescribed intervals. Most of the sporadic action statements prepared for insects are long overdue for revision. Indeed, more generally, many action plans or recovery plans are ineffective as not implemented or inadequately funded, and remain largely as 'desk exercises' that satisfy some formal requirement, and with their conservation guidance and recommendations not translated into practicality.

In some cases, a wider perspective of a taxonomic group enables selection of the most deserving (generally, most threatened) species by some wider consensus, and evaluation against the IUCN criteria (p. 46) or some functional parallel to these. In preparing Australia's national Action Plan for butterflies (Sands and New 2002), advice was solicited from as many informed enthusiasts as possible, including those with direct field experience and interests in the fauna, and catered for regional peculiarities by holding a series of nine workshops covering all States and mainland Territories to give the broadest possible national consensus on levels of threat to all taxa and also to discuss ambiguities and the problems of status assessment and promoting interest more widely. An attempt to determine which Australian butterflies are now the most susceptible to loss (Geyle et al. 2021) distinguished 26 taxa especially prone to extinction—again from consensus of experienced lepidopterists, many of them hobbyists with substantial field experience far exceeding that of many 'professional scientists'. However, even for some butterflies, substantial gaps in knowledge persist and gaining such consensus over some taxa is tenuous, with widely differing opinions held.

Odonata are the only other order amenable to relatively comprehensive evaluation, as discussed by Hawking (1999). For all other groups, conservation advocacy flows largely from the plight of individual species, both listed and more informally noted, or of associations of species in particular environments. Victoria, for example, has an informal Advisory List of Threatened Invertebrates, mostly those of signaled concern, needing further investigation and as possible future listing candidates.

Table 9.1 The seven main priorities for advancing insect conservation in Australia suggested by New (1984)

1. A broad education programme to emphasise the importance of insects
2. Increasing consideration, perhaps through legislation, of environmental changes on insects, together with monitoring impacts
3. Detecting and listing insect species and groups (such as endemic radiations) that merit priority consideration for conservation, perhaps as the basis for a local Red Data Book or similar directory
4. Accelerate progress toward comprehensive habitat representation in reserves, including establishment of specific reserves for insects
5. Increase funding and other support for production of field guides and synoptic works on Australian insects
6. Progressively, fund autecological studies of more notable species as basis for conservation management
7. Recognition that simply 'listing' a species is not alone sufficient for conservation, and that prohibition of take, unless justified soundly, may also have harmful effects

When the list was last revised, in 2009, the 102 insect species listed included many regarded as Data Deficient and for which more information was needed.

Wide consultation and communication is a key to establishing conservation status and accumulating biological and distributional understanding of all such species, and the recent trend toward encouraging responsible 'citizen science' is a major contributor to such endeavour. Wide communication was included in an early list of priority themes for advancing insect conservation in Australia (New 1984, Table 9.1) that was initiated in the hope of stimulating wider involvement and interest. All those points are still highly pertinent pointers to future need, and the importance of advancing scientific knowledge and awareness of conservation need within an increasingly bioliterate society remains critical and urgent.

9.2 Increasing the Profile of Insect Conservation

Education and awareness underpin the future community capability to conserve insects, and fostering those interests is a vital component of insect conservation endeavour. Symbioses between scientists and citizenry have been essential in many such cases in Australia, and have considerable potential to augment activities of conservation-orientated non-government organisations. Advice from the entomologist fraternity (such as through the Australian Entomological Society's Conservation Committee) on any notable species, or wider entomological relevance, may aid many community efforts and interests. Many of the themes noted above endorse the widespread sentiment that invertebrate biodiversity conservation needs a far higher public and political profile than it has received in the past (Hochkirch 2016). His suggestion of high-level investment in one or more centres for invertebrate conservation to accelerate and focus the needed documentation by describing

Table 9.2 An eight-point strategy needed for closing the gaps inherent in data deficiency (Hochkirch et al. 2020) (clear targets for each are needed, together with effective and measurable indicators)

1. Revive explorative field research to redress decline in basic biological and distributional information
2. Link taxonomy information to conservation information
3. Improve global collation of spatial biodiversity data
4. Map spatial threat data
5. Automate pre-assessments for red listing, based on known distribution and existing threats within range
6. Facilitate knowledge transfer, such as bridging gaps between scientists and citizenry
7. Create biodiversity monitoring programmes for lesser-known taxa
8. Provide funding mechanisms to fill knowledge gaps

species, defining their distributions, clarifying their ecology and vulnerability to threats, and assessing their conservation status—so, leading to effective practical conservation management and priorities—would do much to advance those needs. Although this ambition is perhaps largely fanciful, Hochkirch noted that some parallels for other fields of conservation exist. A key need would be for effective liaison with practical managers and ecosystem stewards to 'encourage and guide action-building capacity and raise awareness of the global and national importance of invertebrates'. Providing the knowledge and resources needed by managers to protect invertebrates is vital.

Conservation status assessments of insects, notably 'Red Listing' involving appraisal against the IUCN criteria or some agreed equivalent that can help to overcome major difficulties (such as assessing future population size trends quantitatively), are perennially frustrated by lack of taxonomic and ecological information as well as inability to produce or estimate quantitative threshold levels other than in very general terms. The Golden sun-moth (*Synemon plana*, p. 200) has now been studied quite extensively, as a flagship species for native grasslands in the southeast. An attempt to use Bayesian analysis to explore its conservation status (Mata et al. 2016) produced highly disparate opinions and rankings as the discussion participants were obliged to make categoric decisions from, in some cases, very limited experience of the moth's complex biology. It is almost inevitable that a range of experiences and awareness will be reflected any such outcome. The problems are recognised widely, and many remedial measures have been advocated. Addressing the broad themes of data deficiency is integral to assessments of conservation status and needs. A key component of the eight-point strategy outlined by Hochkirch et al. (2020) (Table 9.2) is 'building capacity in taxonomy, ecology and biodiversity monitoring in countries with high species richness and endemism', as emphasised earlier (Hochkirch 2016). For most insects and other invertebrates both the fundamental data needed and the ability to interpret it are lacking, and those 'lacks' include experts, funding, public awareness and political will. One ramification is that current red lists and their parallels inevitably assess numerous insects as 'Data

Table 9.3 The various categories and criteria for justifying species' assessment as 'Data Deficient', as classified by IUCN (as listed by Bland et al. 2016)

Justification	Definition
Type series	Species known from one collection event at one locality, from which a name-bearing specimen or specimens have been designated
Few records	Species known from five or fewer records
Old records	Species known from records collected prior to 1970
Uncertain provenance	Species known from specimens with very uncertain locality information, therefore species' distribution cannot be mapped
Taxonomic uncertainty	Species for which uncertainty regarding taxonomy directly leads to paucity of data on distribution, population status, ecology, and threats
New species	Species described in the last 10 years before assessment
Uncertain population status or distribution	Species for which information on geographical distribution, population size, and population trends is unavailable or uncertain
Uncertain threats	Species for which information on threats (and species' response to those) cannot be determined

Deficient', a situation that can help to emphasise the magnitude of further work needed.

Knowledge gaps and consequent status uncertainty are greatest in those regions with highest insect richness and/or endemism, and where expertise to investigate most insect taxa is low. Formal 'Data Deficient' (DD) status is often used as a refuge allocation, obliged by the lack of sound evidence to place a species more clearly into any better-defined category of risk. Many may indeed be at risk, but the array of contexts is often difficult to appraise. Bland et al. (2016) identified eight main justifications for applying 'Data Deficient', in pointing out that such species can contribute high uncertainty in evaluating extinction risk across taxonomic groups and that many global projections for conservation that rely on knowledge of threatened species exclude those listed as 'DD'. Despite a clear definition of DD as reflecting 'inadequate information to make a direct, or indirect, assessment of its risk of extinction based on its distribution and/or population status', a later recommendation from IUCN (2017) has been to use the category 'only when there is really no alternative'. Bland et al. held that fuller consideration of the causes of any individual DD allocation (Table 9.3) may help to determine priorities and to direct conservation actions by use of 'justification tags' for each designated species. There may also be considerable differences between taxonomic groups, leading to inconsistent rankings. For most invertebrates, it is inevitable that many higher groups will remain sufficiently unknown that any species-level considerations of conservation status will be highly speculative, and also difficult to communicate to parties beyond the few specialist enthusiasts to whom the group is of interest. For some, emphasis on their ecological roles may accord them some notice and support, but the multitude of taxa for which even functional importance is obscure or not recognised widely—the 'meek inheritors' noted by New (2000)—are passengers under wider

Table 9.4 The main paragraph headings of the Australian Statement on Invertebrates (ConCom 1989), as pointers to summarise the importance of invertebrates in easily-understood terms and emphasise the need for their conservation

1. Invertebrates are the largest component of our fauna, both in number of species and biomass
2. Invertebrates are an important source of food for other animals
3. Invertebrates may also constitute a source of food for mankind
4. Invertebrates are vital to nutrient recycling and to the fertilisation and growth of the vast majority of cultivated plants
5. Invertebrates are useful in protecting farming, forestry, animal husbandry, human health and water purity
6. Invertebrates are valuable aids for medicine, industry and crafts
7. Many invertebrates are attractive
8. Some invertebrates may harm human activities, but their populations may be controlled naturally by other invertebrates
9. Mankind can benefit greatly from enhanced knowledge of invertebrates
10. Terrestrial, aquatic and aerial invertebrates should be protected from possible causes of damage, impairment or destruction

conservation measures. Formal investigation and status assessment for most of these, beyond tacit grouping as Data Deficient, is highly unlikely to occur.

The inter-related themes of increasing awareness, education, cooperation, and logistic support are all key to advancing insect conservation in Australia. All are based in needs for sound documentation of the fauna, realistic appreciation of the threats it faces, and how these may be alleviated, centred in the importance of conserving even unknown insects in a rapidly changing world. A range of contexts for furthering awareness and education have yet to be realised fully. Boppre and Vane-Wright (2012) argued that effective education to increase awareness of insect conservation is a major role of butterfly houses and other institutions featuring live insect exhibits—but considered that in global terms this was only rarely realised. Melbourne Zoo's lead role in conservation of the Lord Howe Island stick insect (p. 200) through captive breeding and informative exhibition is a leading example in Australia.

However idealistic the themes listed in Table 9.1 may seem, their essence has been reiterated in various forms. Australia's 'Statement on Invertebrates' (ConCom 1989), for example, emphasised their importance and predominance in many ecosystems. That Statement was based on the earlier 'European Charter for Invertebrates' and enlarged on the topics listed in Table 9.4, as a justification for invertebrate significance in many walks of human life.

Nevertheless, much of the optimism expressed in that document has not penetrated widely into conservation planning. Athough interests in insect (invertebrate) conservation have indeed broadened, the knowledge and support needed to incorporate them properly into wider conservation agendas in Australia is still manifestly inadequate. The approaches discussed by Taylor et al. (2018) suggest some constructive avenues for advance.

The need to transform taxonomic knowledge of Australia's biota (together with that for New Zealand) is recognised widely (AAS/RSTA 2018), fundamentally to document the nation's biodiversity but with massive benefits in focusing conservation priorities. Six key initiatives have been nominated, and it remains to be found whether the series of broad strategies listed by AAS/RSTA can be accomplished. The initiatives are (1) accelerating discovery of biodiversity; (2) enhancing services for end-users of taxonomic knowledge; (3) engaging with Indigenous knowledge to ensure their interests are included; (4) improving infrastructure by integrating into supportive applications; (5) education for the future to assure recognition and appreciation; and (6) supporting the taxonomic sector by strategic enhancements. Collectively, these themes embrace the stated broad aims to support, enhance and expand taxonomy and biosystematics in the region.

Simply accelerating discovery, as the provision and awareness of the variety present, itself faces formidable logistic barriers. Additional surveys and interpretation are needed badly, but are difficult to accomplish. For any comparative studies across space or time, standardised sampling is clearly needed, and methods used should preferably be straightforward to understand and follow in order to facilitate participation by volunteers. The substantial array of insect sampling techniques (Samways et al. 2010) includes several approaches used commonly in surveys for conservation, but each has numerous variations that, without further standardisation and understanding of limitations, may confound the outcomes. In attempting to bring greater order to this critical need for standardised monitoring insect surveys, Montgomery et al. (2021) recommended standards for seven widely used sampling methods spanning both 'passive' and 'active' approaches (Samways et al. 2010), namely Malaise trapping, light-trapping, pan-trapping, pitfall trapping, beating, acoustic monitoring, and visual surveys such as transect walks. Each can contribute to long-term monitoring, in association with well-defined metadata that can allow better interpretation of changes. Considerable and carefully planned metadata is needed for optimal surveys, and much of the detail suggested by Montgomery et al. is not available for most published surveys. For each of the seven sampling approaches reviewed, the categories of this background information needed are summarised in Table 9.5.

More than one method may be used in a single survey, to target different taxa and complement each other in broader surveys.

Potential conservation management of any insect species prompted by purported decline or threat almost always necessitates some form of standardised monitoring to suggest (1) whether management is needed; (2) when to initiate or modify management; and (3) to assess the outcomes (optimistically, benefits) of management by changes at a site or across sites. Valid monitoring depends on sufficient biological knowledge of the target species to interpret the information gained and avoid both 'false positives' and 'false negatives' that can follow from invalid interpretation. The salutary example of the Golden sun-moth (p. 200) in which detections of incidence and estimates of population size were for long based on adult sightings from visual surveys but without appreciating the strong limits to flight imposed by time of day and influences of weather conditions on moth behaviour provided some misleading

Table 9.5 The recommended background information for insect monitoring programmes: list of required metadata amalgamated for each of seven main methods (Malaise sampling, light trapping, pan trapping, pitfall trapping, beating, acoustic monitoring, active visual surveys) (abbreviated from Montgomery et al. 2021; used under Creative Commons Attribution [CCBY] licence: https://creativecommons.org/licences/by/4.0/)

Locality	GPS coordinates of sampling locations; location description; photo of trap in situ
Site description	Habitat description; photos in four cardinal points, showing habitat; description of plant phenology; amount of light pollution; sampling substrate; substrate size; substrate condition
Temporal	Date trap or monitoring established; date of data collection; time beginning data collection; duration of data collection; time of detection
Environmental	Wind during sampling (Beaufort scale); temperature during sampling; precipitation during sampling; humidity during sampling; cloud cover during sampling; lunar phase during sampling
Sampling description and placement	Trap or sampling equipment type; trap or equipment photo; trap or equipment manufacturer and model; trap or equipment dimensions; mesh hole size, density and shape; killing agent; use of scent or bait; trap orientation; bulb type, wavelength, power and brightness; amount of liquid evaporation during sampling; trap, pan or sheet colour; collecting method (such as aspiration); number of hits per substrate (beating); object dimensions and weight used for hitting; detection distance

outcomes is perhaps not unusual. Small differences in temperature, for example, can markedly affect the activity (and, hence, apparency) of many insects, and designing a robust and reliable monitoring process can only be achieved with that awareness and validation. As another example, the nationally endangered flightless endemic New Zealand grasshopper *Brachaspis robustus* (Acrididae) is highly cryptic in its habitats, occurring on gravels and stony substrates. It is restricted to a small part of the South Island where its habitats have been modified considerably by hydroelectric development and introduced species. The grasshopper's crypsis renders it very difficult to monitor (Schori et al. 2020). Detection depends on movement, and even the large (up to about 3.6 cm body length) females can remain undetected by remaining still (as in cold or overcast weather) and hiding under rocks.

Much material collected in invertebrate surveys is tangential to the primary aim of that survey, and such 'bycatch' is generally not examined fully, and is in many cases discarded rather than archived for future investigation. Their potential as ecological collections (that is, of material collected and archived responsibly as reference points for comparison with future surveys, a principle advanced by Danks et al. 1987) is not then realised, but this loss is only one aspect of the far wider implications of inadequate logistic capability to properly investigate insect diversity. Adequate replication in trapping surveys must be balanced against the reality that many of the insect species encountered are likely to be rare, and that many of the methods cited above kill them. Those surveys may provide both ethical dilemma and additional threat to vulnerable taxa, perhaps necessitating restraint and careful evaluation of the purpose of the survey and the use to which the catches will be put. This inadvertent mortality of sensitive or threatened species is far different from the deliberate targeted take of rare species for commercial gain (including sales on overseas 'black markets') or collector demand, an activity that is commonly prohibited for those species that are formally 'listed' for conservation.

In some cases, although extraordinarily difficult to prove, suspicions arise that the monitoring itself may be the cause of the changes documented. Impacts of light trapping for nocturnal insect sampling, and for long a standard monitoring tool for moths, may become a deleterious contribution to ALAN, for example (p. 85). This suggestion has been investigated for bees in North America, for which killed samples from pan traps and netting were used to examine bee richness and abundance every 2 weeks through the flight season and across several years (Gezon et al. 2015), so that large numbers of bees were removed each sampling day. Interpretations from the 132 morphospecies represented implied that the sampling did not significantly affect the bee assemblages. However, and especially when sampling occurs at short intervals at the same sites and over extended periods, insect diversity and community composition could be influenced by the continuing removal of individuals. The above outcome for bees, essentially lack of any deleterious sampling effect on a key group of pollinators, could reflect several aspects of bee population dynamics and life histories. Gezon et al. (2015) discussed four such possibilities: (1) compensatory mortality and natality, whereby removal of bees may reduce competition for resources and increase population growth rates; (2) sampling might remove individuals from the population that would not have reproduced, so functionally distinguishing 'effective population' from 'census population'; (3) many solitary bees may be sampled after they have mated and provisioned brood cells (as a continuing process throughout their life), so that reproduction could mitigate sampling losses; similarly, sampling of workers or post-mated males of eusocial bees may have minimal effects on reproduction; and (4) high species turnover associated with short flight seasons of many species, so that new species or populations are being sampled on each sampling occasion.

These variations emphasise that definitive insect inventories, even for a limited range of taxa, are complex to compile and that short-term surveys—as the most usual practice dictated by logistics and opportunity—can fall far short of total evaluation, notwithstanding the valuable data they yield. The British experiences with long-term insect monitoring based on standardised replicable sampling techniques and efforts, such as the programmes that led to robust concerns over moth declines (p. 14), are exceptional. However, long-term and short-term monitoring programmes can give very different outcomes, with the former influenced also by any site/habitat changes (such as by succession) as the period increases, but with species accumulation curves for either approach rarely reaching an asymptote. For European Odonata, only 53% of species were captured in 1 year of survey. This rose to 65% in three consecutive years of sampling, and 16 years were needed to reach 95% (Dolny et al. 2021). The more elusive species included some of higher conservation interest as rare and specialised taxa. Assessing the vagaries of cross-site inventories from short-term (1–2 years/seasons) surveys may not be sufficiently reliable for founding conservation measures and must be balanced carefully against the impracticalities of gaining more reliable data from longer investigation. Dolny et al. suggested need for sequential sampling over at least 10 years (yielding, in their odonate study, >80% of species), with this period sometimes taken as the low threshold for defining 'long-term' monitoring in ecological studies. They cautioned, also, that the purpose of any such exercise must be defined clearly at its commencement.

Surveys are perhaps particularly urgent in vulnerable ecosystems such as alpine areas, as ecosystems subject to likely imminent influences from climate changes, and in which insect diversity is both under-documented and has substantial levels of endemism. As a parallel, surveys of alpine beetles in New Zealand (Paler et al. 2020) by using pitfall traps in *Chionochloa* grasslands revealed that Coleoptera were diverse. Many species were captured in very low numbers and appeared to be rare. Up to 60% of the 139 species/morphospecies were represented by five or fewer individuals, and 28% of species only by singletons. That pattern of distribution in species' abundance (with many species found only in very low numbers) is not unusual across many insect habitats, but emphasises that a substantial proportion of species present may be elusive, some perhaps genuinely rare, and difficult to study and assess properly. As discussed by Novotny and Basset (2000), based on their extensive samples of phytophagous insects from a series of host trees in Papua New Guinea in which singletons were numerous (180 of 347 chewing species from *Ficus* and Euphorbiaceae, and 98 of 444 sucking species on *Ficus*), interpreting what these apparently low abundance species truly represent is difficult. Five scenarios were projected for this plant-related context, as (1) species that are transient, present casually as 'tourists' and not feeding on that plant; (2) species that are sampled inadequately by the techniques employed, so that their appearance is an artefact of the methods used; (3) specialists with genuinely low population levels; (4) generalist species feeding occasionally on the plant on which they were found but with much higher populations if assessed across their full host range; and (5) specialists feeding on the sampled plant but which prefer, and are more abundant on, a few other hosts, usually closely related, as oligophagous feeders. Of these, perhaps only the third category is a genuine conservation concern, as 'genuinely rare' and possibly vulnerable. However, in diverse plant communities (lowland tropical rainforest in Novotny and Basset's study), the numbers of transient species can become inflated easily by 'spillover' from nearby plants, and their abundance is simply accidental. Even after these were excluded, numerous singletons remained, and the last two of the above categories were the most important sources of the rare species encountered.

For highly endemic fauna and vulnerable environments, many such genuinely low abundance taxa may disappear unheralded as climates and other conditions change. The Australian alpine insect fauna is likely to be equally vulnerable and also includes numerous locally endemic taxa that render comparisons even across geographically close mountain areas difficult.

9.3 Citizen Science in Australian Insect Conservation

In the last few decades, increased participation by 'citizen scientists' has contributed significantly to numerous insect surveys and conservation monitoring in Australia and augmented the limited professional expertise and leadership available. Citizen scientists generate the great majority of terrestrial biodiversity observations and records (Groom et al. 2017) and, in a broad sense, a citizen scientist 'is a volunteer who collects and/or processes data' (Silverton 2009). Volunteer involvement in such exercises is by no means limited to insects, of course, but the principle of partnering dedicated volunteer enthusiasm and natural history interests with scientific leadership is both a means of countering the 'extinction of experience' so justifiably lamented by Cheesman and Key (2007) and providing vital field information. Those involvements are the backbone of many conservation activities. The contexts for volunteer involvements in insect conservation vary greatly, spanning recording or monitoring incidence of individual species to more extensive surveys of sites to appraise wider diversity. Initiated well before the term 'citizen science' became popular, the contributions from large numbers of volunteers are the foundation of insect recording activities promoted by organisations such as Butterfly Conservation (United Kingdom, Warren 2012), Buglife (United Kingdom: Stubbs and Shardlow 2012) and the Xerces Society (North America: Black 2012) and progressively emulated elsewhere. In Australia, networks of volunteers, some based in 'Friends Groups', have been instrumental players in recording and monitoring notable threatened insects such as the Richmond birdwing butterfly (*Ornithoptera richmondia*) in Queensland and New South Wales (Sands and New 2013) and the Eltham copper butterfly (*Paralucia pyrodiscus lucida*) in Victoria (New 2011)—these, and some others have become long-term continuing exercises that could not continue effectively without that community participation. Citizen science projects formed through 'interest groups' can prove very rewarding, as the above have demonstrated, and can sometimes be formed rapidly in response to focused need. An appeal through the Entomological Society of Victoria (Anon 2020) drew attention to the unexpectedly high diversity of the tiny Sun-loving moths (Heliozelidae) associated with flowers of native Rutaceae. More than 100 new moth species have been found associated with only about 50 plant species (that is, about 10% of the total native Rutaceae species), and the appeal seeks help in progressively increasing this representation by photographing moths on flowers of other Rutaceae. As in many similar projects, social media are important components of reporting outcomes.

The nature of participation in wider field surveys is exemplified by surveys for the larger beetles of Lord Howe Island and the nearby small Blackburn Island, New South Wales (Reid and Hutton 2019; Reid et al. 2020), organised through the Australian Museum and Lord Howe Museum. The major purpose was to survey the beetles present before an eradication campaign to eliminate rats from Lord Howe was commenced, so to set a baseline against which to evaluate changes, with an aim also to instruct volunteers in survey techniques. Focus on the larger beetles (>1 cm body length) restricted the scope of the survey and facilitated non-specialist recognition of the targets. Volunteers were given an initial presentation on Lord Howe beetles and supplied with information sheets on the larger-bodied taxa present. Searches were led by two experienced scientists in the field, one of whom was on hand to appraise beetles as they were discovered in each of the 13 habitats/sites inspected. At least 100 species of beetles were collected in the initial investigation over 22–28 July 2018. Most of the (about) 535 beetle species recorded from the Lord Howe group are endemic and, as elsewhere, many are small, but 23 of the approximately 75 'large' species were collected by the citizen scientist surveys in 2018 and 2019. Volunteers found several species not found by the leaders. The most notable finding was the endemic flightless *Cormodes darwini* (Cleridae) on Blackburn Island, as a species that had been considered extinct since its last recorded incidence in 1916, shortly before predatory rats arrived on Lord Howe in 1918. *Promethis sterrha* (Tenebrionidae), another flightless endemic, was also found on rat-free Blackburn Island, and several other discoveries were notable additions to their previous single or few records.

A major initiative that continues to contribute substantially to survey of protected areas (p. 124) and documentation of Australia's biota through combining citizen and scientist interests is 'Bush Blitz', launched in 2010 as a partnership led by the Australian government and also involving BHP Billiton, Earthwatch and the National Scientific Reference Site Network. The principle is to use teams of scientists and volunteers to explore a designated site (or series of sites) over several days of concentrated collecting and observations in order to document their fauna and flora more extensively. The specimens collected and photographs contributed to iNaturalist (p. 179), and thence the Atlas of Living Australia, continue to add considerably to basic documentation The almost 40 expeditions completed up to 2019 included all States and Territories (Fig. 9.1), and all have yielded putative new species (Bush Blitz 2020), with numbers reflecting ability to recognise these. Among insects, the greatest number of 'new' species (462) is for true bugs (Hemiptera: Heteroptera), reflecting their wide distribution and that a core group of taxonomists has been available to assess captures and records. Indeed, the only 'new' species found on one expedition, to small islands in the Coral Sea, were two species of mirid bugs. Heteroptera are followed by moths (149 putative new species), jumping plant lice (128) and bees (112) with lower numbers of several other orders. Shortage of taxonomic expertise and resources has hampered fuller appraisal of many groups, and is a major impediment to wider insect documentation in Australia. Primary focal groups for each Bush Blitz survey differed with the interests of survey participants, and the individual illustrated reports for each expedition contain much original information on the environments and their biota.

Bush Blitz surveys, even as time-restricted 'snapshots' of insects present, have provided important new records and information for many taxa. For example, freshly collected material of small parasitoid wasps, in condition suitable for DNA analyses, has facilitated greater understanding of diversity and relationships. Thus, Fagan-Jeffries et al. (2019) acknowledged the surveys as instrumental in obtaining fresh material from remote locations and described new microgastrine wasps (Braconidae, one of them given the species name '*bushblitz*') based on survey material. Similarly,

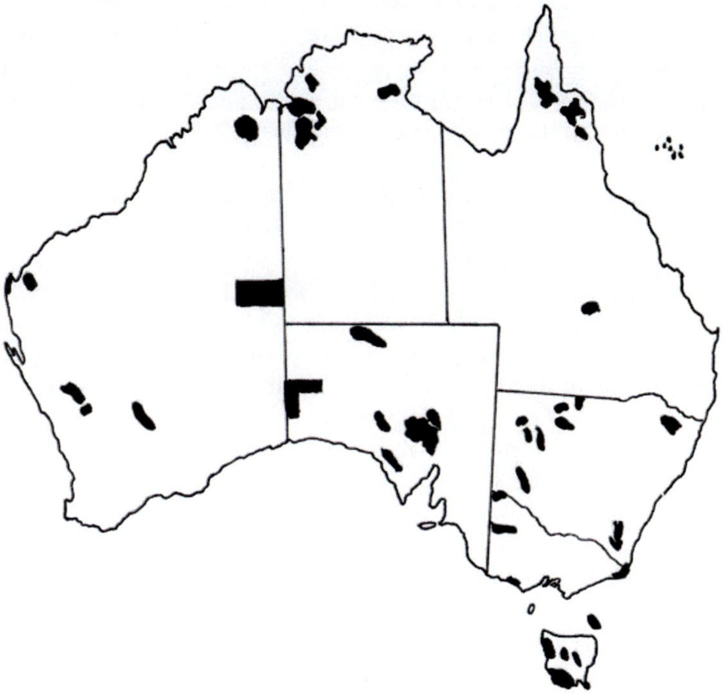

Fig. 9.1 The distribution of Bio Blitz surveys across Australia up to 2018. Shaded areas are sites (some multiple) where one or more surveys have occurred (from Bio Blitz 2020)

descriptions of new cheloninc wasps (Braconidae) incorporated material from three arid zone Bush Blitz surveys (all on reserves that were previously sheep-grazing stations) in South Australia (Kittel and Austin 2016).

Ideally, both volunteers and science benefit from these interactions, and notable discoveries such as the above create lasting memories. The scientific and communication values of citizen science collaborations in many parts of the world have been discussed extensively (Kosmala et al. 2016). Success depends on clear recognition of the purpose of an exercise, and equally clear recognition of any focal taxa, together with constructive advice and feedback as guidance and to reduce errors. Thus, for surveys of ladybird beetles (Coccinellidae) in Europe and North America, volunteer recorders had access to on-line identification guides for these often colourful beetles, and sightings/records were submitted through digital photographs for validation (Gardiner et al. 2012). Caution is clearly needed when 'look-alike' species occur and could be confused. A few reports of the Eltham copper butterfly (above), for example, were of the similar but more widespread species with rather similar biology in Victoria, the Bright copper (*Paralucia aurifer*). For ladybirds, Gardiner et al. noted possible confusions within polymorphic species, but also other sources of error, including lack of experience of some volunteers, effectiveness of training and, for more detailed surveys, complexity of sampling protocols. The latter becomes highly relevant when quantitative data, such as abundance or monitoring changes, are sought, rather than 'presence/absence' or simple detection information, and credibility of identifications is a prelude to dissemination or publication of the results that, in Gardiner et al.'s words aid in 'documenting the contributions of thousands of amateur scientists' and assuring that those contributions match the standards of those from 'professional scientists'. Projects involving citizen scientists need very careful planning and direction, with clear design and replication of any sampling protocols together with instruction to ensure they are used consistently and as intended, and procedures for validation and assessments of accuracy (Kosmala et al. 2016). Particularly for focal species of conservation concern, misidentifications—either 'false positives' or 'false negatives'—could influence status assessment and practical consequences. Many insects are not sufficiently distinctive to reward citizen science surveys, although the spectrum of focal candidates continues to increase, so that incorporation of ecological assessments by photographic recording of pollinators (as a group of major ecological concern and importance) is sometimes feasible, and has been undertaken in several places. Advice from relevant specialists is clearly necessary to assure accuracy of citizen records of many taxa, and validation through voucher specimens or photographs recommended. Large, distinctive, diurnal conspicuous insects—commonly those acknowledged by people and for which some form of field guide may be available to facilitate recognition—are amongst the most tractable targets. Butterflies and dragonflies are prime candidates in Australia.

The United Kingdom Pollinator Monitoring Scheme aims to find out how insect pollinator populations are changing across Great Britain, using standard survey forms and identification guides and incorporating the expertise of several collaborating partner organisations and volunteer networks: full details are given on the scheme's website. That scheme focuses on bees and hoverflies, because they provide a high proportion of 'pollination services' to both crops and wildflowers. Two levels of counts are undertaken: (1) flower-insect timed counts, the total number of insects visiting a particular flower species in a 10-min period in warm dry weather over April–September, and (2) 1 km square counts, using pan traps to sample a wider array of insects over, collectively, a series of 75 such squares across the UK.

The consequences of widespread concern for pollinators in North America have for long been a leading activity of the Xerces Society (Mader et al. 2011), continuing through projects such as the 'Pollinator Protection Policy' by which people are asked to adopt the four key principles of (1) growing a variety of bee-friendly plants that flower from spring to autumn; (2) protect and provide bee nests and caterpillar host plants; (3) avoid use of toxic pesticides, especially insecticides; and (4) talking about the importance of pollinators and their habitats—all activities in which people can participate easily. Declines of bees, in particular, continue to arouse wide concerns and attention (Winfree 2010; Winfree et al. 2011) and high levels of declines are repeatedly reported. Even for Europe, more than half the native bee species are too data-deficient for reliable evaluation of their conservation status (Nieto et al. 2014). For North America, 1437 of the total 4337 native bee species were sufficiently well-known for evaluation and more than half of these (749: 52%) were declining and 24% threatened with extinction (Kopek and Burd 2017). Many others, poorly known species, are also likely to be at risk, and many of the detected threatened bees were endemic, some with very narrow distributions. More broadly, more than 40% of global pollinators are 'highly threatened' (Potts et al. 2016). Major threats to North American bees were llnked to agricultural intensification, with losses of natural habitats, extensive crop monocultures and use of toxic pesticides. Some bee species were associated clearly with particular flower species for pollen: the Californian vernal pool systems in which Yellow carpet (*Blennosperma nanum*, Asterales) grows are vital for the monolectic Yellow carpet solitary bee, *Andrena blennospermatis*. Up to 90% of that habitat has been lost (Kopek and Burd 2017). Climate change and urbanisation were also identified as major threats.

A consideration highlighted recently (Iwasaki and Hogendoorn 2021) is that much community concern for pollinators is derived from concerns over crop pollination services associated mainly with a few widespread bee species (notably *Apis mellifera*) that have strong industry-based advocacy to create awareness of conservation need. However, in Australia (and some other countries) *A. mellifera* is not truly a native species—although naturalised after some 200 years of occupation—and measures to enhance such predominant generalist taxa may not benefit many more specialised native pollinators. For these, a perspective that extends well beyond needs for crop pollination to include native flora and their associates is necessary.

Australia's 'Wild Pollinator Count' surveys, designed primarily for parts of New South Wales and Victoria, accumulate records of a variety of pollinating insects based on surveys of single plant species by observers, with identifications facilitated by an illustrated brochure of many of the most likely candidates, and opportunity to submit photographs and comments. Surveys are taken over designated time periods, a week each in spring and late summer, with each observer selecting the target plant species for observation at some time in that period. Such broad interests can extend easily to education and the involvement of young people in insect recording and observations. School-age children can contribute information and, at the same time, be encouraged to increase their awareness of insect life through simple experiments on school grounds. For example, the 'Pollinators in the Playground' project used observations of insect visitors to artificial flowers of different colours (Prendergast et al. 2021), drawing from the broader suggestions for school-based pollinator projects (Saunders et al. 2018). Any such avenues to fostering interests among young people are significant investment for the future. An earlier example was the participation of numerous schools in New South Wales and Queensland in rearing and recording the Richmond birdwing butterfly (Sands and New 2013, p. 204). In both these examples, a major benefit was liaison with experienced scientists as advisors and 'mentors'—an activity that can help to foster interests beyond the immediate project itself. A more recent example, is the partnership ('Insect Investigators') between taxonomists and several schools in South Australia through which Malaise traps monitored by pupils on or near school grounds allow students to participate actively in the naming of new species and learning about the processes involved. Fagan-Jeffries et al. (2021) included the unusual authorship attribution of 'and Citizen Science Participants of Insect Investigators' to acknowledge this

Table 9.6 Ten principles of Citizen Science, as adopted by the Australian Citizen Science Association (abbreviated from ACSA 2020)

1. Citizen science projects actively involve citizens in scientific endeavour that generates new knowledge or understanding
2. Citizen science projects have a genuine science outcome
3. Citizen science provides benefits to both science and society
4. Citizen scientists may participate in various stages of the scientific process
5. Citizen scientists receive feedback from the project
6. Citizen science, as with all forms of scientific inquiry, has limitations and biases that should be considered and controlled for
7. Where possible and suitable, project data and meta-data from citizen science projects are made publicly available and results are published in open access format
8. Citizen scientists are suitably acknowledged by projects
9. Citizen science programmes offer a range of benefits and outcomes which should be acknowledged and considered in project evaluation
10. The leaders of citizen science projects take into consideration legal and ethical considerations of the projects

cooperation and in describing four new species of braconid wasps Fagan-Jeffries and Austin (in the above paper) noted potential for expanding the programme as a means to promote wider community connections with taxonomists.

The potential for proliferation of citizen science projects for insect conservation in Australia is limited only by the imagination of the proposers, and their ability to excite public interest and participation and to sustain a 'critical mass' of interest. Projects can occur in all environments, including urban area surveys, where the pool of potential participants may be large and lack of costs and time in reaching more remote areas be a substantial logistic advantage. In Australia a series of key principles for the success of citizen science projects has been proposed (Table 9.6, ACSA 2020), emphasising the wide variety of cooperations and responsibilities that arise. Several other recently-established entomological citizen scientist campaigns Australia focus on popular taxa—the 'Butterflies Australia' phone app/web site seeks to increase records of butterflies throughout Australia, and its launch in 2019 was celebrated by issue of a postage stamp depicting an endemic butterfly in a commemorative stamp series acknowledging citizen science projects. This ambitious project seeks to integrate citizen science with standardised monitoring protocols and the developing technological and social media tools to rapidly increase the data available on Australian butterflies and, hence, to facilitate monitoring of distributions and abundance through time: its details are discussed by Sanderson et al. (2021). A primary aim is to overcome the barriers imposed by data deficiency for numerous taxa that are at present poorly known, and four broad general aims for Butterflies Australia are listed by Sanderson et al., as '(1) to raise community awareness of butterflies and invertebrates in general, specifically the need for more research and conservation effort; (2) to encourage community participation through good user interface design in the app and website, including provision of a free field guide for the app; (3) to increase availability of scientifically robust observational data on butterflies; and (4) to influence policymakers to provide more attention to butterflies and potentially other invertebrates in their available funding and policy design'. A laudable long-term aim is to replicate the outcomes achieved for the North American eButterfly project, for which eButterfly (2020) reported more than 410,000 occurrence records contributing to a fuller picture of that fauna since the scheme commenced in 2012.

A more specific project, also launched in 2019, is 'Moth Tracker', designed to allow people to report incidences of the Bogong moth (*Agrotis infusa*, p. 22) during its migrations in spring and when, in the past, vast numbers of moths moved from lowlands to aestivate in caves and rock shelters in the Australian alps. Recognising that attraction to artificial lights may have contributed to the moth's recent decline, an allied campaign 'Lights off for the Bogong moth' calls for switching off of unnecessary external lighting along the main migration routes, in attempts to reduce mortality. This campaign was initiated through concerns not primarily for the moth, but through the lessening numbers of the endangered Mountain pygmy possum (*Burramys parvus*), for which the aestivating moths are a major food resource during the breeding season. Indeed, the dependence of the possum on moths ensures that support for the moth becomes an integral component of the more widely-advocated values of conserving *Burramys*. It exemplifies the wider potential for purposefully including insect conservation in wider agendas in Australia. Deliberate promotion of insect conservation through such 'piggy-backing' could become far more widespread: although disparaging evaluations of insects as 'bird food' and the like diminish their importance to many people, they also emphasise their central roles in many food webs, and such roles, when defined, provide strongly positive conservation motivation. Some early accounts of the Bogong moth reported large numbers of ravens emerging from caves and feeding on aestivating moths (Warrant et al. 2016), and suggested that the birds may have helped people to locate those sites for their own benefit.

Development of on-line recording platforms for animals and plants, supported wherever possible by photographic records of sightings has led to major advances in documenting biodiversity. They range from taxon-limited coverage to very broad accumulations of records. The global platform iNaturalist is a major concentration of citizen scientist records, for which Mesaglio and Callaghan (2021) noted that Australian records (across all taxa) include more than 36,000 species represented by more than 1.6 million observations, and contribute to the Atlas of Living Australia database. Arthropods dominated animal records, with insects comprising 30.5% of all Australian observations and 33.6% of all taxa. As might be expected, the Lepidoptera were the most recorded insect group, with 50.4% of all insect observations. However, numerous other groups are also represented, with records of some threatened species considerably augmenting earlier knowledge. iNaturalist incorporates a wide range of constituent projects—for example, Mesaglio and Callaghan noted 'Cicadas of Australia', with more than 10,000 observations across 166 species and—of considerable significance—the 'First Known Photographs of Living Species' project that included pictures of 183 Australian insect species for which no previous images were available. For some species, those recent

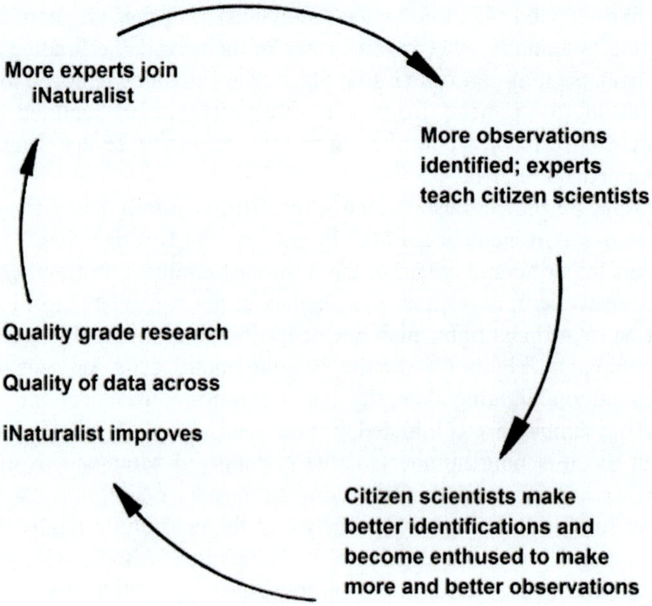

Fig. 9.2 The 'iNaturalist loop' displaying the projected positive feedback as iNaturalist grows in Australia: as more taxonomic experts participate, more observations will increase the reliability of the data and lead to more data for ecological and conservation research questions (after Mesaglio and Callaghan 2021)

photographs constitute the first records since they were originally described, and the records thereby have considerable taxonomic importance. The potential of iNaturalist (and other such platforms) for conservation flows from this increased documentation and that the records submitted can be scrutinised (and in many cases the subject identified reliably) by specialists, who can then post comments on significance and conservation status. The 'feedback loop' created by those interactions (Fig. 9.2) illustrates how ecological and conservation research may be enhanced by increased participation by taxonomists continuing to increase the quality and interest of the data. A related conservation theme is provisional information on possible insect-plant interactions. Many photographs of bees, hoverflies and butterflies were of the insect on a flower, suggesting possible ecological associations for investigation (Gazdic and Groom 2019).

Mesaglio and Callaghan (2021) foreshadowed the greater integration of Australian iNaturalist data into professional ecological research as the value of the information is progressively recognised, this also likely to increase the uses to which it may be put.

Discussion of a small group of tiny flightless endemic pygmy grasshoppers (Tetrigidae), all of them very rare, exemplifies the revelations possible from such social media records (Skejo et al. 2020) and for which those records of species restricted to the eastern rainforests added substantially to the limited knowledge previously available. Eight of the 11 new records reported were on-line. One species (the Angled Australian barkhopper, *Paraselina brunneri*) had not been recorded since it was described in 1887 (although a more recently described species [1952] was here synonymised with it). Similarly, *Paraselina trituberculata* (the Triple-bump Australian barkhopper) had not been seen since it was described in 1932 and the photograph reproduced was of the first known male of this distinctive species. Records of such rare (or, at least, elusive) species confound implications of their extinction and reduce levels of data deficiency.

Appreciation of the value and significance of the work undertaken is a key factor in citizen science or wider public involvement, and consensus among the variety of stakeholder groups can endorse acceptability in education and outreach measures. Public understanding avoids misinformation that could hamper or impair management plans for declining species and more general biodiversity conservation. The example of assessing public knowledge and understanding of bees in Canada (van Vierssen Trip et al. 2020) is, perhaps, not atypical. A series of 2000 respondents to telephone interviews suggested that public knowledge of bees was low, although many people considered that bees should be protected, predominantly for their 'ecological services'. However, awareness of native bee diversity was rather low, with many respondents identifying the European honeybee as of paramount importance. This perception was paralleled in a survey of media coverage of pollinators in Australia (Smith and Saunders 2016) in which (from 151 items over 2006–2015), *Apis mellifera* comprised half the total coverage, native bees featured in only 17% and non-bee pollinators mentioned in only 17%. This led Smith and Saunders to suggest that such over-simplistic appreciation of diversity might be a more widespread perception. For bee conservation, in principle interest can be high, but the level of understanding is often only low (Wilson et al. 2017): for many insects of less tangible significance or benefits to humanity bridging that awareness gap is a formidable task.

However, the uses to which the records and accumulated data flowing from citizen science projects are put are not always clear. People may have concerns over indiscriminate use of records of threatened species, for fear of encouraging exploitation, and also of commercial use of the data they provide. Opinions on open access, whereby the data become readily available for use in honing research and in conservation planning, differ widely (Fox et al. 2019). Surveys in the United Kingdom sought opinions on open access to information from two major Lepidoptera recording schemes ('Butterflies for the New Millenium', 'National Moth Recording Scheme'), and revealed considerable reluctance to open or unrestricted use, with strong feeling (79.6% of 104 regional recorders) that this should be limited to non-commercial uses. A possibly general concern is that citizen scientists want their own data and records to contribute to conservation, but fear that this aim may become undermined. Fox et al. exemplified this fear within the contexts of

(1) collectors targeting rare species; (2) deliberate destruction of habitat by landowners to avoid conservation restrictions and responsibility; and (3) accidental damage to the sites by people wanting to see scarce species. Related concerns include that release of fine-scale information could exacerbate this and that a time delay (some respondents suggesting at least 5 years) may be warranted. For threatened species, many respondents recommended some 'blurring' of the information released, with release of fine-scale information receiving little support in relation to progressive acceptance at scales of 1 × 1, 2 × 2 and 10 × 10 km squares. For widespread species this reluctance largely disappeared, but some concerns over immediate data access remained.

Reflecting human population distribution in Australia, citizen science exercises are proving particularly valuable in and near urban areas. They are major contributions toward wider acceptance of insects as 'worthy' and overcoming the adverse public perceptions that have hampered much progress in the past. Fostering positive perceptions of insects is a universal need in insect conservation and, as for conservation itself, can occur across a hierarchy of levels, from individual to global (Fig. 9.3, Duffus et al. 2021) to enable more effective and publicly-accepted conservation to occur.

9.4 Flagship Species

Publicity in support of, or garnering awareness of, any notable insect species, can be instrumental in promoting conservation. The captive breeding programme that was key to preventing extinction of the Lord Howe Island stick insect (*Dryococelus australis*, p. 200) (Honan 2008) led to well-informed public display at the Melbourne Zoo. A recent fund-raising campaign for conservation of the Green carpenter bee (*Xylocopa aerata*) on Kangaroo Island (Hogendoorn et al. 2021, p. 204) increased awareness of the intricacies of fire management for insects, and the damage that can occur to sensitive species. Icon species with public appeal are exemplified by Leichardt's grasshopper (*Petasida ephippigera*), one of Australia's most colourful Orthoptera and endemic to small parts of the Northern Territory, including Kakadu National Park, where it is promoted as of interest to tourists. One of about 160 species of grasshoppers in the park (Andersen et al. 2001), *P. ephippigera* is restricted to sandstone heath communities supporting its food plants (*Pityrodia* spp., Lamiaceae). Populations are influenced by fire (Barrow 2009) and *P. ephippigera* has considerable cultural significance to the local Jawoyn people (Andersen et al. 2014) so strong local support for its conservation, largely through managing fire regimes, is evident. The grasshopper is pictured widely on tourist literature and souvenirs.

In Queensland, the Ulysses butterfly (*Papilio ulysses*, Papilionidae) is a parallel 'tourist icon' pictured heavily on souvenir goods and regional publicity documents. It was listed for protection, somewhat controversially, in the mid 1970s with expressed concerns over harm to this spectacular species from overcollecting. This situation was doubted strongly, and the listing occurred without consultation with

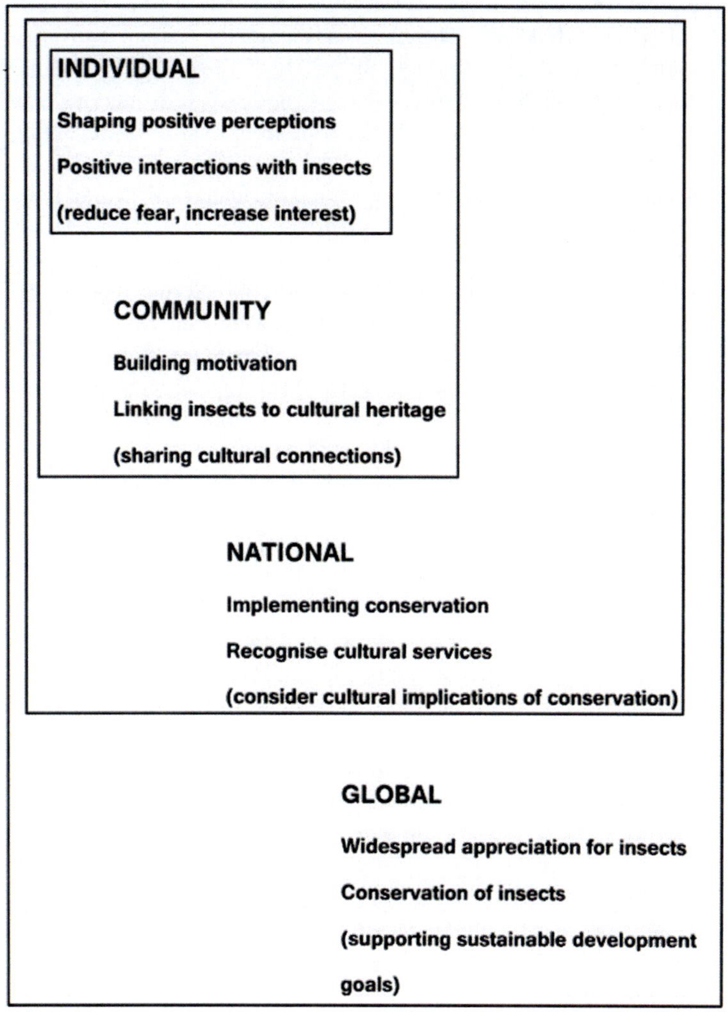

Fig. 9.3 The multi-level framework for increasing positive perceptions of insects (based on Duffus et al. 2021, used under Creative Commons Attribution [CCBY] licence: https://creativecommons.org/licences/by/4.0/)

the knowledgable local entomological community and without sound evidence. Widespread belief that this listing was not justified on grounds of conservation need led to considerable reluctance to pursue this approach in the state and hampered constructive progress in insect conservation there. Such attractive appearance can promote wider interests. The spectacular Atlas moth of north west Australia (*Attacus wardi*, Saturniidae) is one of the largest moths in the country, with female wingspan reaching up to about 19 cm. After its discovery in the early twentieth century, it was not seen again for some 70 years but is now known from several widely separated sites (Braby and Nielsen 2011), although seemingly extinct at its type locality, Darwin, due to urbanisation. *A. wardi* is projected as a flagship for conservation of coastal tropical monsoon forest communities, where larvae feed on *Croton habrophyllus* (Euphorbiaceae). The moth is apparently restricted to larger forest patches (a minimum of 8 ha, but preferably >20 ha: Lane et al. 2010) with adequate *Croton*, but most of this forest type now consists of small disjunct patches within savanna woodland. Braby and Nielsen noted the importance of preventing further loss of this critical habitat, eradication/control of grassy weeds and reducing the impacts of late dry season fires in the savanna, all features that may threaten this scarce moth, which is still very poorly known.

Co-occurrence of spectacular threatened species can augment the conservation significance of sites. In Queensland, Mary Cairncross Scenic Reserve, Malveny, supports populations of both the Richmond birdwing butterfly (p. 204) and the endangered Southern Pink underwing moth (*Phyllodes imperialis smithersi*, Erebidae), confined to small remnant patches of subtropical rainforest in which the endemic larval foodplant vine grows (Sands 2012), emphasising the importance of such small isolated areas for endemic flora and insects.

Iconic or popular insects are not necessarily threatened species. For example, the enormous mounds created by some termites are popular sightings on some tourist routes in northern Australia. Those of the Cathedral termite (*Nasutitermes triodiae*) are a major feature of Kakadu National Park (Andersen et al. 2014), and tourist brochures for Litchfeld National Park (Northern Territory) proclaim that those mounds are 'one of the park's most impressive sights'. Mounds are commonly up to 4 m high, but can occasionally attain heights of 8 m, and support vast numbers of individual termites. The large numbers of generally somewhat smaller mounds constructed by magnetic termites (*Amitermes meridionalis*) can still each contain a million or more termites, and mounds sometimes occur in large groups that, owing to their uniform orientation, have been likened to graveyards (Korb 2003). These grass-feeding termites, thus, are among the most abundant insects in the region, and are also among the most ecologically dominant insects in the northern Australian savannas. These termite mounds are used as nesting sites by some rare parrots that, in turn, harbour specific moths unknown to breed elsewhere (p. 20).

Any insect species—especially those without direct pest, health or adverse welfare connotations—noticed by people can become useful for education and promoting awareness of issues related to insect values, roles and conservation. Unusual appearances, behaviours and biology are so frequent among insects that it can require little imagination to fire public interest by an enthusiastic advocate, with the message embracing the needs for conservation and the idiosyncrasies of each species or other kind of insect. 'Rarity', as a more general attribute of value to many people, can also be appreciated widely, especially in fostering local interests within a circumscribed range. However they are defined, 'rare species' are considered widely to be intrinsically more prone to extinction than are non-rare species (Fattorini et al. 2013), so that rarity becomes an indication of extinction risk. Both the Eltham copper and Richmond birdwing butterflies (above) were appreciated because of geographical rarity and vulnerability together with unusual (or novel) biological features that caught public interest—for *Paralucia*, the tripartite association between butterfly, specific ant and specific foodplant, and the nocturnal activity of caterpillars led to interest and volunteers participating in surveys and site maintenance and restoration, and for *Ornithoptera* the 'ecological trap' of an alien food plant attracting oviposition but not enabling larval development was a key point and stimulus for removal of the vine and widespread planting of the native food plant as a substitute to benefit this spectacular butterfly. Parallels could be extended to almost every individual species for which conservation needs have been identified.

Only in very rare cases is such publicity of doubtful value, and is exemplified by the publicising of some rare butterflies from novel sites, from which they might be exploited by commercial or over-zealous collectors, in seeking to reduce any threat from this process. Whilst 'prohibition of take' is a widespread condition in conservation legislation, collector intensity in Australia is generally rather low and over-collecting was not considered a widespread threat by Sands and New (2002). However, fear of prosecution for capturing specimens and the difficulties of obtaining legitimate permits for study (even when not 'handling' specimens) was a widespread deterrent to advancing butterfly interests in Australia, alienating many of the people who could contribute most to additional information on rare species with which they were familiar, for fears of prosecution for pursuing their hobby interest. For some Australian butterflies, notably among some complexes of Hesperiidae and Lycaenidae, incidences, including range extensions, can be confirmed only by examination of voucher specimens, with even good quality photographs of some species not revealing the detail required to confirm identity. Imposed inability to capture and retain individuals, and thwarting sincere efforts to add to fundamental knowledge of notable species can erode interest, or drive it 'underground'. Especially for butterflies, but also for some other 'collectable' insects such as jewel beetles (Buprestidae), 'collector lore' is a rich potential source of information, much of which never enters the formal record. The series of regional workshops held during compilation of the national Butterfly Action Plan (Sands and New 2002), for example, sought that wide participation, and the information proferred and discussed considerably augmented the credibility of the final document. Concerns over 'black market' take by and for overseas collectors have extended to some endemic stag beetles (Lucanidae), jewel beetles (Buprestidae) and spectacular dragonflies, notably the petaltails, Petaluridae, all of which are very localised and of conservation concern. Indeed, there is suggestion that there is considerably more material of tropical *Petalura* in overseas collections than in Australian collections, leading to calls for greater control of their export (Hawking and Theischinger 2004), however difficult that may be.

The diversity of citizen science contributions to insect conservation links also with the importance of individual actions. Kawahara et al. (2021) discussed eight such 'simple actions' that can be undertaken by people without any specialised knowledge or training and that may rapidly improve conditions for insects and increase awareness of insect needs and importance. The first five of these (Table 9.7) are focused on habitat improvements, and the last three are orientated more to increasing public appreciation of insects and the needs to conserve them. In such ways, awareness and 'bioliteracy' can become widespread, as Janzen and Hallwachs (2021) have advocated for the Costa Rica 'BioAlfa' programme whereby

Table 9.7 Individual's contributions to saving insects from decline: the eight actions proposed by Kawahara et al. (2021), compiled from text paragraph headings

Create insect-friendly habitats:
1. Convert lawns into diverse natural habitats
2. Grow native plants
3. Reduce pesticide and herbicide use
4. Limit use of exterior lighting
5. Lessen soap run-off from washing vehicles and building exteriors, and reduce use of driveway sealants and de-icing salts
Increase awareness and appreciation of insects:
6. Counter negative perceptions of insects
7. Become an educator, ambassador, and advocate for insect conservation
8. Get involved in local politics, support science and vote

'all members of a country can attain some level of proficiency among citizen scientists' and as the first such enterprise for a tropical country. Both the principle and the actions are largely familiar in Australia, with some knowledge of the unique features of the biota widespread among both First Nations and the more recent populace: the main need is to enhance awareness of insects and other invertebrates within this scenario. The citizen science projects noted above are valuable contributions to that endeavour.

References

AAS/RSTA (Australian Academy of Science/Royal Society Te Aparangi) (2018) Discovering biodiversity: a decadal plan for taxonomy and biosystematics in Australia and New Zealand, 2018-2027. Canberra, Wellington

ACSA (Australian Citizen Science Association) (2020) Obtainable. https://citizenscience.org.au/who-we-are/. Accessed 28 Mar 2021

Andersen AN, Ludwig JA, Lowe LM, Rentz DCF (2001) Grasshopper diversity and bioindicators in Australian tropical savannas: responses to disturbance in Kakadu National Park. Austral Ecol 26:213–222

Andersen AN, Humphrey C, Braby MF (2014) Threatened invertebrates in Kakadu National Park. In: Widerlich S, Woinarski J (eds) Symposium 7: conservation of threatened species, 26–27 March 2013, internal report 623. Supervising Scientist, Darwin, pp 48–57

Anon (2020) Insect and plant lovers – we need your help in a citizen science project !!!! Vict Entomol 50:106

Barrow P (2009) Role of fire in the ecology of Leichardt's grasshopper (*Petasida ephippigera*) and its foodplants, *Pityrodia* spp. Ph D Thesis, Charles Darwin University, Darwin

Black SH (2012) Insect conservation and the endangered species act: a history. In: New TR (ed) Insect conservation: past, present and prospects. Springer, Dordrecht, pp 171–189

Bland LM, Bielby J, Kearney S, Orme CDL, Watson JEM, Collen B (2016) Toward reassessing data-deficient species. Conserv Biol 31:531–539

Boppre M, Vane-Wright RI (2012) The Butterfly House Industry: conservation risks and education opportunities. Conserv Soc 10:285–303

Braby MF, Nielsen J (2011) Review of the conservation status of the Atlas Moth, *Attacus wardi* Rothschild, 1910 (Lepidoptera: Saturniidae) from Australia. J Insect Conserv 15:603–608

Bush Blitz (2020) Statistics. https://bushblitz.org.au/statistics/. Accessed 4 July 2021

Cheesman OD, Key RS (2007) The extinction of experience: a threat to insect conservation? In: Stewart AJA, New TR, Lewis OT (eds) Insect conservation biology. CABI, Wallingford, pp 322–348

ConCom (Council of Nature Conservation Ministers) (1989) Australian statement on invertebrates, Canberra

Danks HV, Wiggins GB, Rosenburg DM (1987) Ecological collections and long-term monitoring. Bull Entomol Soc Canada 19:16–18

Dolny A, Pyszko P, Sigutova H (2021) Community changes in odonate monitoring: why are long-term studies so relevant? Insect Conserv Divers. https://doi.org/10.1111/icad.12491

Duffus NE, Christie CR, Morimoto J (2021) Insect cultural services: how insects have changed our lives and how can we do better for them. Insects 12:377. https://doi.org/10.3390/imsects12050377

eButterfly (2020) eButterfly: an online database of butterfly distribution and abundance [web application]. Available at http://www.e-butterfly.org (accessed July 2021)

Fagan-Jeffries EP, Cooper SJB, Austin AD (2019) New species of Australian microgastrine parasitoid wasps (Hymenoptera: Braconidae: Microgastrinae) documented through the 'bush Blitz' surveys of national reserves. Zootaxa 4560(3):401–440

Fagan-Jeffries EP, Austin AD, Citizen Science Participants of Insect Investigators (2021) Four new species of parasitoid wasp (Hymenoptera: Braconidae) described through a citizen science partnership with schools in regional South Australia. Zootaxa 4949(1):79–101

Fattorini S, Di Giulio A, Dapporto L (2013) Measuring insect rarity: practical issues, pragmatic approaches. J Insect Biodivers 1(10):1–21

Fox R, Bourn NAD, Dennis EB, Heafield RT, Maclean IMD, Wilson RJ (2019) Opinions of citizen scientists on open access to UK butterfly and moth occurrence data. Biodivers Conserv 28: 3321–3341

Gardiner MM, Allee LI, Brown PMJ, Losey JE, Roy HE, Smyth RR (2012) Lessons from lady beetles: accuracy of monitoring data from US and UK citizen-science programs. Front Ecol Environ 2012. https://doi.org/10.18890/110185

Gazdic M, Groom Q (2019) iNaturalist is an unexploited source of plant-insect interaction data. Biodivers Inform Sci Stand 3:e37303

Geyle HM, Braby MF, Andren M, Beaver EP, Bell P et al (2021) Butterflies on the brink: identifying the Australian butterflies (Lepidoptera) most at risk of extinction. Austral Entomol 60:98–110

Gezon ZJ, Wyman ES, Ascher JS, Inouye DW, Irwin RE (2015) The effect of repeated, lethal sampling on wild bee abundance and diversity. Methods Ecol Evol 6:1044–1054

Groom Q, Weatherdon L, Geijzenorffer IR (2017) Is citizen science an open science in the case of biodiversity observations? J Appl Ecol 54:612–617

Hawking JH (1999) An evaluation of the current conservation status of Australian dragonflies (Odonata). In: Ponder W, Lunney D (eds) The other 99%. The conservation and biodiversity of invertebrates. Royal Zoological Society of New South Wales, Mosman, pp 354–360

Hawking JH, Theischinger G (2004) Critical species of Odonata in Australia. Int J Odonat 7:113–132

Hochkirch A (2016) The insect crisis we can't ignore. Nature 539:141

Hochkirch A, Samways MJ, Gerlach J, Bohm M, Williams P et al (2020) A strategy for the next decade to address data deficiency in neglected biodiversity. Conserv Biol 35:502–509

Hogendoorn K, Glatz RV, Leijs R (2021) Conservation management of the green carpenter bee *Xylocopa aerata* (Hymenoptera: Apidae) through provision of artificial nesting substrate. Austral Entomol 60:82–88

Honan P (2008) Notes on the biology, captive management and conservation status of the Lord Howe Island stick insect (*Dryococelus australis*) (Phasmatodea). J Insect Conserv 12:399–413

IUCN (2017) Guidelines for using the IUCN red list categories and criteria. Version 13, IUCN, Gland

Iwasaki JM, Hogendoorn K (2021) How protection of honeybees can help and hinder bee conservation. Curr Opin Insect Sci 46. https://doi.org/10.1016/j.cois.2021.05.005

Janzen DH, Hallwachs W (2021) To us insectometers, it is clear that insect decline in our Costa Rican tropics is real, so let's be kind to the survivors. Proc Natl Acad Sci 118:e2002546117

Kawahara AY, Reeves LE, Barber JR, Black SH (2021) Eight simple actions that individuals can take to save insects from global declines. Proc Natl Acad Sci 116(2):e2002547117

Kittel RN, Austin AD (2016) New species of Australian arid zone chelonine wasps from the genera *Phanerotoma* and *Ascogaster* (Hymenoptera: Braconidae) informed by the 'Bush Blitz' surveys of national reserves. J Nat Hist 50:211–262

Kopek K, Burd LA (2017) Pollinators in Peril. A systematic status review of North American and Hawaiian native bees. https://biologicaldiversity.org/campaigns/native_pollinators/pdfs/Pollinators_in_Peril.pdf

Korb J (2003) The shape of compass termite mounds and its biological significance. Insect Soc 50: 218–221

Kosmala M, Wiggins A, Swanson A, Simmons B (2016) Assessing data quality in citizen science. Front Ecol Environ 14:551–560

Lane DA, Martin G, Weir RP (2010) The life history of *Attacus wardi* Rothschild (Lepidoptera: Saturniidae) from the Northern Territory, Australia. Austral Entomol 37:115–127

Mader E, Shepherd M, Vaughan M, Black S (2011) Attracting native pollinators. Protecting North America's bees and butterflies. Storey Publishing, North Adams, MA

Mata L, Garrard GE, Kutt AS, Wintle BC, Chee YE et al (2016) Eliciting and integrating expert knowledge to assess the viability of the critically endangered golden sun-moth *Synemon plana*. Austral Ecol 42:297–308

Mesaglio T, Callaghan CT (2021) An overview of the history, current contributions and future outlook of iNaturalist in Australia. Wildl Res. https://doi.org/10.1071/WR20154

Montgomery GA, Belitz MW, Guralnick RP, Tingley MW (2021) Standards and best practices for monitoring and benchmarking insects. Front Ecol Evol 8:579193. https://doi.org/10.3389/fevo.2020.579193

New TR (1984) Insect conservation – an Australian perspective. W Junk, The Hague

New TR (2000) How to conserve the 'meek inheritors'. J Insect Conserv 4:151–152

New TR (2011) Butterfly conservation in South-Eastern Australia: progress and prospects. Springer, Dordrecht

Nieto A, Roberts SPM, Kemp J, Rasmont P, Kuhlmann M et al (2014) European red list of bees. European Union, Luxembourg

Novotny V, Basset Y (2000) Rare species in communities of tropical insect herbivores: pondering the mystery of singletons. Oikos 89:564–572

Paler K, Monks A, Leschen RAB, Ward DF (2020) Determining species diversity and functional traits of beetles for monitoring the effects of environmental change in the New Zealand alpine zone. Ecol Indic 121:107100. https://doi.org/10.1016/j.ecolind.2020.107100

Potts SG, Imperatriz-Fonseca I, Ngo HT, Aizen MA, Biesmeijer JC et al (2016) Safeguarding pollinators and their values to human well-being. Nature 540:220–229

Prendergast K, Vanderstock A, Neilly H, Ross C, Pirotta V, Tegart P (2021) Potential and pitfalls of citizen science with children: reflections on Pollinators in the Playground project. Austral Ecol. https://doi.org/10.1111/aec.13031

Reid CAM, Hutton I (2019) Citizen science and the art of discovery: new records of large Coleoptera from Lord Howe Island, July 2018. Tech Rep Aust Mus 28:1–16

Reid CAM, Hutton I, Thompson S (2020) The citizen scientist survey of large Coleoptera on Lord Howe Island, August 2019. Tech Rep Aust Mus 31:1–15

Samways MA, McGeoch MA, New TR (2010) Insect conservation. A handbook of approaches and methods. Oxford University Press, Oxford

Sanderson C, Braby MF, Bond S (2021) Butterflies Australia: a national citizen science database for monitoring changes in the distribution and abundance of Australian butterflies. Austral Entomol 60:111–127

Sands DPA (2012) Review of Australian *Phyllodes imperialis* Druce (Lepidoptera: Erebidae) with description of a new subspecies from subtropical Australia. Austral Entomol 39:281–292

Sands DPA, New TR (2002) The action plan for Australian butterflies. Environment Australia, Canberra

Sands DPA, New TR (2013) Conservation of the Richmond birdwing butterfly in Australia. Springer, Dordrecht

Saunders ME, Roger E, Geary WL, Meredith F, Welbourne DJ et al (2018) Citizen science in schools: engaging students in research on urban habitat for pollinators. Austral Ecol 43:635–642

Schori JC, Steeves TE, Murray TJ (2020) Designing monitoring protocols to measure population trends of threatened insects: a case study of the cryptic, flightless grasshopper *Brachaspis robustus*. PLoS One 15(9):e0238636. https://doi.org/10.1371/journal.pone.0238636

Silverton J (2009) A new dawn for citizen science. Trends Ecol Evol 24:467–471

Skejo J, Connors M, Hendriksen M, Lambert N, Chong G et al (2020) Online social media tells a story of *Anaselina*, *Paraselina*, and *Selivinga* (Orthoptera, Tetrigidae), rare Australian pygmy grasshoppers. ZooKeys 948:107–119

Smith TJ, Saunders ME (2016) Honey bees: the queens of mass media, despite minority rule among insect pollinators. Insect Conserv Divers 9:384–390

Stubbs A, Shardlow M (2012) The development of Buglife- the invertebrate conservation trust. In: New TR (ed) Insect conservation: past, present and prospects. Springer, Dordrecht, pp 75–105

Taylor GS, Braby MF, Moir ML, Harvey MS, Sands DPA et al (2018) Strategic national approach for improving the conservation management of insects and allied invertebrates in Australia. Austral Entomol 57:124–149

van Vierssen TN, MacPhail VJ, Colla SR, Olivastri B (2020) Examining the public's awareness of bee (Hymenoptera: Apoidae: Anthophila) conservation in Canada. Conserv Sci Pract 2:e293

Warrant E, Frost B, Green K, Mourutsen H, Dreyer D et al (2016) The Australian Bogong moth *Agrotis infusa*: a long-distance nocturnal navigator. Front Behav Neurosci 10:77. https://doi.org/10.3389/fnbeh.2016.00077

Warren M (2012) Butterfly conservation: the development of a pioneering charity. In: New TR (ed) Insect conservation: past, present and prospects. Springer, Dordrecht, pp 133–154

Wilson JS, Forister ML, Carril OM (2017) Interest exceeds understanding in public support of bee conservation. Front Ecol Environ 15:460–466

Winfree R (2010) The conservation and restoration of wild bees. Ann N Y Acad Sci 1195:169–197

Winfree R, Bartoneus I, Cariveau DP (2011) Native pollinators in anthropogenic habitats. Annu Rev Entomol 42:1–22

Chapter 10
A Future for Australia's Insects

10.1 Introduction

Much of the impetus for the processes and involvements discussed in the last chapter depends on people valuing insects as part of the unique heritage and natural environments of Australia, and appreciating that they are indeed unique, irreplaceable and worth conserving. Recognition of insect diversity, and that much of that diversity—from individual species to entire communities—is composed of species found nowhere else in the world and is vulnerable to an array of anthropogenic disturbances, is a key to emphasising and promoting conservation based on the best information available, but also with recognition that this information may be highly incomplete. Developing policies to achieve this effectively, and translating those policies and priorities urgently into action for the best possible outcomes demands strong national commitment, with leadership and participation by the Commonwealth and agencies of all States and Territories. Within that broad ambit, setting priorities for allocation of limited funding and logistic expertise must be seen to be both responsible and credible, and policy and management integrated carefully. Those activities and symbioses must also apply to the entire range of relevant insect habitats and ecosystems, from natural vegetation and waterbodies at present little changed to highly altered and disturbed agricultural, industrial and urban landscapes and waterbodies that need to be variously protected and promoted through restoration, reservation or enhancement of connectivity and natural resources that can enable insects to thrive, where necessary colonise, and persist. A wide landscape-scale perspective may be needed, rather than focusing solely on small sites or remnant habitats alone. For highly mobile insects such as the Richmond birdwing butterfly (*Ornithoptera richmondia*, p. 204) broad scale landscape enrichment by plantings of the native larval food plant vine was instrumental in allowing the butterflies to naturally expand distribution to cover much of their historical range in Queensland and New South Wales, and in reducing levels of threat (Sands and New 2013).

© The Author(s), under exclusive license to Springer Nature Switzerland AG 2022

T. R. New, *Insect Diversity, Declines and Conservation in Australia*, Fascinating Life Sciences, https://doi.org/10.1007/978-3-030-90134-9_10

Setting optimal priorities can reveal the major gaps in information needed on topics including conservation status, threat intensity and variety, habitat needs and management priorities. Any insect conservation programme must remain flexible and sufficiently adaptable to respond to new information, and so requires effective monitoring against specified target criteria. Almost all 'species management plans' include pleas for more research and elucidation of particular aspects of the species' ecology, distribution, and responses to change. Continuing investigations of Australian insects, with an overarching long-term aim of enumerating and diagnosing the taxa present, can progressively clarify both taxonomic and ecological complexity and the extent of loss of sensitive habitats and sites and, so, the vulnerability of numerous local endemic species. Increasing formal attention to insects (and other invertebrates) in pre-development 'environmental impact statements' is perhaps unlikely to occur widely, other than for the few 'listed species', as is increased long-term monitoring of development impacts on insects on the changed sites and adjacent affected areas.

Some designated threatened species were brought to initial attention principally because of their scarcity, apparent restriction to single sites or specialised habitats seen to be vulnerable, and only after listing have more comprehensive appraisals of biology and conservation status ensued. In many cases this has occurred only because of some 'official interest' obliging that investigation. The Eltham copper butterfly (p. 202) came to the fore in Victoria after wide belief that it was extinct was corrected by discovery of a small population on a site scheduled for imminent housing development. That early history (New 2011) reflected it becoming an important flagship for both insect conservation and the importance of urban remnant environments for insects, as well as stimulus for additional surveys and study through the aid of directed citizen science groups, and led also to designation of some of the State's first dedicated butterfly reserves. Such progress and continued involvement remains unusual and, for even a small range of individual listed species—and notwithstanding formal obligations to investigate further—is often disregarded. Victoria's Flora and Fauna Guarantee Act, for example, obliges preparation of an Action Statement for each listed species, but most that have been prepared for insects are long past their revision date, and unfulfilled. 'Conservation of threatened species', as one of the eight general 'no-regret measures' for insect diversity conservation suggested by Harvey et al. (2020) (Table 4.1, p. 70), is a mainstay of insect conservation as perceived by the public and is central to increasing awareness and interest. However, it has ramifications across all other measures listed as evaluating threats and countering their influences: all measures are relevant in Australia, but simply listing them—whilst demonstrating the variety of needs—also masks the formidable complexities involved in each, and defining the relative impacts of the threatening processes detected.

For each of the major stressors for insects in Australia, conservation draws on some general principles that are almost universal, but often difficult to evaluate fully and to remedy in particular contexts. Many proposals thereby appear idealistic and impracticable, but each principle is also an important consideration in optimising the balance between the options available. Very particular responses may be needed for localised species and/or small sites for which local conditions and threats may provide unique suites of concerns. In contrast, a broader approach to conserving insect diversity across wider arenas may draw effectively and directly on principles of landscape ecology. Those principles have been iterated frequently, and are exemplified by the six listed by Samways (2020) as guides to measures likely to have wide benefits to incompletely defined biodiversity. Most have very wide relevance, as: maintaining protected areas and extending their extent and quality; maintaining high habitat quality and heterogeneity; reducing hard boundaries and contrasts between landscape elements, so providing a greater range of microclimates and buffers across edges; increasing functional connectivity by conservation corridors and habitat 'stepping stones'; and reducing the alien features of novel and transformed landscapes. Other general measures overlap variously with these but can emphasise more local contexts. Thus, the policy solutions suggested for European butterflies (Warren et al. 2021) included needs to maintain traditional management of key habitats in conjunction with agriculture and forestry activities, so avoiding intensification as a prime concern in butterfly declines. Samways' last pointer, simulating natural disturbances, emphasises the dynamic nature of many insect habitats and that management to maintain particular successional stages or conditions suitable for critical resources may need to heed the kinds of disturbances that are not anthropogenic. Modifying imposed burning treatments in Australia to more closely resemble the low intensity mosaic burns of pre-European wild fires rather than the massive areas subject to control burns in forests at present, for example, may have substantial benefits in enhancing connectivity and providing refuges. Management of agricultural and urban/periurban areas can markedly influence their capability to host native insects, and the principles involved are well understood—even if not always heeded. Attention to retention, enhancement and restoration of small remnant natural vegetation patches, and retention of large old trees in urban green spaces, are among the measures recommended for fostering urban biodiversity (Ikin et al. 2015). Likewise, the influences of climate change can be inferred sufficiently to include consideration in long-term planning at the landscape scale to either facilitate natural insect movements or to select and/or protect sites for future colonisation or translocations.

This chapter includes a number of practical suggestions to advance insect conservation in Australia. It is not a 'recipe book', but presents selected examples that demonstrate the ways in which policy, scientific investigation and conservation management interact, and the practical opportunities and constraints involved. The representative cases are each based on concerns for individual species or sites, as the most tangible manifestations of interest. It is presumed, usually without rigorous justification, that parallel declines occur both in distribution and abundance of individual species and the wider richness of insects across all ecosystems they inhabit. That precaution underpins much current concern for invertebrate conservation.

10.2 Integrating Insect Conservation in Australia

10.2.1 Protective Legislation

The differing coverages and obligations of threatened species legislations across Australia were summarised by Taylor et al. (2018). Inconsistencies are not uncommon, so that some EPBC-listed species are not listed under range state legislation (and vice-versa), or different levels of threat are given under different legislations. These differences or omissions can cause occasional tensions or confusion through differing assessments of conservation status between the Commonwealth EPBC Act and regional legislations, and there are further inconsistencies in the relative coverage of threatened communities and threatening processes. Aligning the consequences of listing under the various Australian legislations was regarded as a priority need by Taylor et al. (2018) who advocated that all nominations and listings should be made under the EPBC Act. They also noted the development of a more uniform standard assessment for species pursued under a Memorandum of Understanding between the Commonwealth and States/Territories with the aim of preventing duplication of effort and resources at the two 'levels'. The principle of that memorandum has been agreed by all states and territories, and is intended to promote 'a more consistent, efficient and harmonised process', with the intention that assessments made under other jurisdictions will be adopted under the EPBC Act, using the IUCN Red List categories of threat. Duplications and, importantly, inconsistencies will gradually be eliminated, as all species are assessed on a national scale and species will be listed in the same category across all participating jurisdictions. Some delays are inevitable and reflect the large number of 'legacy species' that have at present been listed in one or more range jurisdictions but not assessed under the proposed common method. Some changes to current evaluations are likely—both in upgrading listing status and in de-listing some taxa—as inconsistencies are investigated. Current 'regional scale assessments' are not compliant with the common method. However, listings of some species not considered nationally threatened may still be pursued by individual regional jurisdictions as state or territory threatened species, and are not included in the EPBC Act list of nationally threatened species.

Historically, 'listing' for invertebrates has followed from examples pioneered from listing vertebrates and, to a somewhat lesser extent, vascular plants. As Kitching pointed out some years ago (Kitching 2004), this species-level approach has worked reasonably well for Australian mammals and birds, due to three major influential factors: (1) the organisms are generally sufficiently large to achieve a public and political presence; (2) the species and the higher taxa involved are generally sufficiently well-known taxonomically that most of the 'problem' species can be accommodated; and (3) there are relatively few species involved. Yet, even for the most 'charismatic' Australian vertebrates, legislative listing for protection can become extraordinarily hard to enforce over large remote areas and across lands with differing tenures and priority uses—including economic activities such as mining, forestry and agriculture with the arrays of threats they may generate to both terrestrial and aquatic environments. The perspectives and prospects are far different for invertebrates. For insects (1) public and political profiles are generally non-existent, in part reflecting the small size of most insects; (2) taxonomy is generally highly incomplete, with numerous species undescribed or, even, undiscovered; and (3) enormous numbers of species are present in most insect orders. In Australia 'only a few families of invertebrates are understood taxonomically to a level comparable with the vertebrates' (Kitching 2004), and for most species biological and distributional knowledge is fragmentary. Nevertheless, there is expectation that listed insects and other invertebrates are as 'tangible' as the better documented and more popular vertebrates, and that detailed conservation measures are equally feasible: in reality, most status assessments or management documents prepared as responses to listing rely heavily on recommendations for general measures such as habitat protection and 'prohibition of take', with only limited detail possible on the biological framework in which practical conservation measures must be undertaken. Effective listing can target only those species that are known well enough for reasonable ecological evaluation. Even in far better-documented faunas, many ecologically significant insect groups are impossible to appraise properly. Thus, considering parasitoid Hymenoptera in Britain 'for most species there is no biological information at all, while for the majority of others the supposed information that has been published is so full of error that the impression of knowledge is extremely misleading' (Shaw and Hochberg 2001). In essence diverse whole trophic guilds such as this are excluded from consideration for species-level conservation treatments. The notion of data deficiency extends well above individual species, with the large knowledge gaps for many Australian insects leading to highly incomplete and taxonomically biased suites of listed species: indeed Kitching (2004) characterised the then representation as 'almost whimsical'—but also 'probably inevitable' because of the massive size of the fauna and few people dealing with them. Victoria's additional 'Advisory list' (p. 164) simply suggests some of the numerous supposedly worthy and priority candidates that may become appraised for formal listing at some later date but which, in the meantime, should be protected wherever possible.

Despite the severe limitations in understanding and possible ambiguities of status, the few listed insect species are the public ambassadors for eliciting wider support and interest in insect conservation, and remain valuable and tangible foci for gaining good will and sympathetic understanding, as long as their credibility can be assured.

10.2.2 Conservation in Practice

To many people, insect conservation is just about single species, brought to notice because they are perceived to be threatened or in some way 'special' or notable, and often occurring on sites of wider interest, for which those species can be used to promote wider protection. Study and conservation of such species has demonstrated both how such operations can be planned and shown to work and the importance of public and political interests in undertaking them. These operations are exemplified below, but increasing realisation that there will always be far more needy species than can be treated individually in such costly and labour-intensive programmes dictates that a wider perspective, encompassing greater insect diversity and the varied ecosystems they frequent, must urgently be recognised as a priority. In all, advances devolve on successful partnerships and cooperations between individuals, local authorities and government agencies.

10.3 Species Conservation: Lessons from Experience

The representative cases of insect species of conservation concern in Australia display the wide variety of contexts, concerns, and gaps in knowledge that have helped to guide development of awareness and capability. The ten cases of endemic insects outlined below represent a range of orders, and are among the rather few that are sufficiently well-documented to discuss how and why they have attracted attention, and collectively indicate the variety of cases and contexts that have drawn attention, and that raise many different issues—all of which have far wider relevance in future insect conservation practice. Most have been discussed more fully elsewhere and, rather than recapitulating their full conservation history, the main points and innovations of each case are noted against a brief background history of interest, together with the major lessons from each. Some are discussed elsewhere in this book, as indicated, and key references to each are given to facilitate further investigation.

Lack of information on some species and wider aspects may be in part compensated by the process of seeking constructive advice from 'experts' toward any consensus over status or needs. The process is sometimes called 'expert elicitation' and may, at least, narrow future options. Recent such exercises on the Golden sun-moth (Mata et al. 2016, p. 200) and for determining which Australian butterflies may be at greatest risk of extinction (Geyle et al. 2021) both illustrate the considerable range of opinions and uncertainties that occur. A wider attempt to use global expert opinion to confirm causes of insect declines and define mitigation strategies included contributions from 413 respondents, from all major regions (Milicic et al. 2021), and emphasised needs for further practical investigations across regions and major insect taxa. Nevertheless, expert opinion provided useful insights into planning and implementation of insect conservation, and suggested some regional differences in relative importance of threats.

10.3.1 *Some Examples*

Odonata

Ancient Greenling Damselfly (*Hemiphlebia mirabilis*) (p. 100)

Listed as Threatened under Victoria's FFG Act and informal assessments as Critically endangered now considered generally unsuitable. Described from a small swamp in central Victoria, this tiny damselfly was for long believed to have been rendered extinct through agricultural conversion of suitable sites in the Yarra Valley of Victoria. It is univoltine, and the weakly-flying adults live for only a few days. As a putative 'living fossil' taxonomically isolated in its own Superfamily (Hemiphlebioidea)—but with this status queried more recently—its re-discovery in the 1980s and conservation became of global priority through the Odonata Specialist Group of IUCN's Species Survival Commission, and helped to draw wider attention to the highly unusual features of Australia's insect fauna. A small population discovered in Wilsons Promontory National Park, Victoria (on a small area then grazed by cattle and susceptible to trampling as the water level lowered in late summer, but with grazing leases now expired), stimulated biological investigations and wider surveys, in course revealing its existence on Flinders Island and in north east Tasmania and, later, very large populations in far south western Victoria and across the border to South Australia. This wider distribution and the very large population (estimated as more than a million individuals) supports earlier opinions that the conservation status should be downgraded from critically endangered, whilst recognising that some isolated small populations on small or ephemeral water bodies may indeed be vulnerable.

Accidental burning of the main Wilsons Promontory site was feared to have extirpated the damselfly there (p. 100), but monitoring showed that it survived, probably through some larvae taking refuge in the deeper, wetter areas previously mown as a firebreak and enabling local recolonisation of the severely burned area over several seasons. Recovery has been aided by recognition of the damselfly's interest by the National Parks staff, who adopted it as one of 'their' special species worthy of conservation. Careful mosaic burning may be a useful management tool to maintain open areas on small occupied swamps as succession occurs.

Useful background references: Cordero-Rivera (2013, 2015), Crowther (2011), Trueman et al. (1992), New (1993).

Giant Petaltail Dragonfly (*Petalura gigantea*)

Listed as Endangered in New South Wales under the Threatened Species Conservation Act. Restricted to areas near the east coast of New South Wales, *P. gigantea* is one of the largest Australian dragonflies, with female wingspan up to about 12.5 cm, and biologically unusual in that long-lived (estimates in the range of 10–30 years for larval development have been suggested) larvae form chambered burrows under swamps and forage on land at night: unlike most Odonata, larvae are not truly aquatic. *P. gigantea* occurs in sphagnum and peaty swamps, and the largest of these, Wingecarribee Swamp, has been exploited for peat mining. A number of historical sites are no longer wetlands, and loss of the dragonfly from much of its historical range appears genuine, reflected in the very conspicuous adult form that is unlikely to avoid notice in targeted searches. Two priority management sites were selected under the NSW 'Save our Species' strategy, and community interest in the dragonfly's conservation may help in future surveys for this spectacular species.

Habitat protection through restrictions on mining and maintaining native vegetation buffers (recommended at 50 m wide, and through liaison with landholders and local authority), with wider rehabilitation of swamp habitats (such as by weed removal and measures to reduce invasion—such as by pines in plantation areas—and measures to reduce erosion and sedimentation) were recommended components of management. Small swamps are vulnerable to local disturbances such as trampling and vehicle damage (such as by fire management vehicles), and impacts of wild fire on areas with peaty soils may be severe.

Useful background references: Baird (2012), Baird and Burgin (2016).

Plecoptera

Mount Donna Buang Wingless Stonefly (*Riekoperla darlingtoni*)

A small flightless narrow range endemic stonefly known only from the summit area of Mt. Donna Buang, within the Yarra Ranges National Park, Victoria, where the main (small) population is in small streams adjacent to a tourist carpark and threatened by runoff (with a comment in the FFG Act Action Statement by Ahern et al. (2003), that 'a single fuel spillage in a carpark could destroy a substantial proportion of the known stonefly population') and expansion/development of tourist activities for winter sports and summer sightseeing. Intensive surveys confirmed that the stonefly is indeed very narrowly distributed, with the only other individuals known discovered about 3 km from the original population. All individuals known are from within about 1 km of the mountain summit, in the narrow elevational range of 1000–1200 m asl. It is unclear whether the stonefly may be susceptible to climate changes as the water warms in its natal streams. Protection of the streams by vegetation buffers is recommended, together with provision for rapid advice and response for any fuel spills or related accidents.

Useful background references: Ahern et al. (2003), New (2008, 2021)

Orthoptera

Key's Matchstick Grasshopper (*Keyacris scurra*)

Listed as Critically endangered under the EPBC Act. A wingless grassland grasshopper endemic to New South Wales, Victoria (where it was for long believed to be extinct, before recent rediscovery and still known only from one locality in the state) and the Australian Capital Territory. *K. scurra*, a member of the endemic subfamily Morabinae, which contains numerous (around 250 species and 41 genera) rather similar flightless taxa, was formerly widespread in grasslands (particularly those dominated by Kangaroo grass, *Themeda triandra*, and with suitable food plants, notably Asteraceae) and grassy woodlands. Expansion of wheat growing and grazing agriculture has reduced these habitats to small remnants, many of them also degraded and unprotected. Concern thus arose from inferred small current range, severe fragmentation of suitable habitat, decline in both habitat extent and quality, and that the processes leading to these situations were continuing. The grasshopper's very limited dispersal capability renders it vulnerable on progressively isolated patches. Several of those refugial areas were in country cemeteries that had for long been fenced to exclude grazing stock and remaining as small islands of native grassland in highly altered landscapes—but which in some cases have now become unsuitable through changed management such as increased mowing and imposed burning regimes. Others included railway easements and travelling stock routes. Recent surveys suggest that the grasshopper might have a rather broader habitat range. Habitat loss and modification is the key cause of decline, and prevention of further losses through protection and management a core of conservation management need. Weed control needs include preventing native trees from colonising selected grassland sites.

K. scurra was the subject of classic studies on chromosomal polymorphisms in the 1960s and so has enduring contributions to understanding Australian insect evolution. It has persisted on some sites since the 1950s and 1960s and also retained high levels of genetic variability over that period—but has also been lost from many historical sites as their condition has changed. However, translocations for 'genetic rescue' are considered unnecessary at present (Hoffmann et al. 2021), and the persistence of genetic variability implies that *K. scurra* may be able to respond evolutionarily to future threats on some sites. More focused protection is warranted for some populations, including the two Victorian populations that are genetically distinct from populations elsewhere and may be the only surviving populations in the state. Protection of their habitat is a critical need.

Rediscovery in Victoria has stimulated wider interest in the grasshopper's conservation, including a five-year plan initiated by Zoos Victoria in 2020 as part of their 'Fighting Extinction' programme. The components of that plan include (1) initiating further surveys in Victoria; (2) supporting habitat reservation and restoration; (3) initiating captive breeding for public display and reintroduction; (4) placing on Zoos Victoria's priority species list; and (5) fostering citizen science for community-led searches.

Useful background references: White (1956), Hoffmann et al. (2021)

Phasmatodea

Lord Howe Island Stick Insect (*Dryococelus australis*) (p. 106)

Listed as Critically endangered under the EPBC Act and Endangered in New South Wales. Known also as the Land lobster, this large wingless phasmid was exterminated from its native Lord Howe Island after rats were introduced in 1918 and believed extinct. Remains found on the steep nearby Balls Pyramid prompted searches on this difficult-to-access islet, and a few individuals were found on an isolated small patch of vegetation, from which two pairs were taken in the hope of founding a captive colony (Priddel et al. 2003). After considerable difficulty (including the female being resurrected from near-death), breeding was established successfully at Melbourne Zoo (Honan 2008). This led to substantial numbers of stick insects being reared over generations, allowing for public exhibition at the zoo (with valuable conservation publicity), foundation of populations in other institutions, and the real prospect of reintroducing the stick insect to Lord Howe. This step awaits confirmation of the success of a rat eradication programme to render that environment 'safe'. It is likely that translocations will initially be made on small islands adjacent to Lord Howe, but it is hoped to eventually reintroduce *D. australis* to the main island.

An unusual and welcome measure has been publication of a childrens' book on the stick insect (Cleave and Tulloch 2015). Targeted at ages 6–9 years, it is accompanied by a series of 'Teachers' Notes' with additional background and a series of questions to promote school class discussions on themes such as 'habitat' and 'endangered species'.

The assessment made as a component of the Lord Howe Island Biodiversity Plan has been adopted as a nationally approved recovery plan for the stick insect, and includes restricting access to Balls Pyramid.

Useful background references: Honan (2008), Priddel et al. (2003), Wilkinson (2014).

Lepidoptera

Golden Sun-Moth (*Synemon plana*) (p. 91)

Formerly listed as Critically endangered under the EPBC Act but status revised (December 2021) to Vulnerable, and Endangered in all three regional jurisdictions across its range. Recorded from native grasslands in New South Wales, Victoria and the Australian Capital Territory, this moth has become an important ambassador for these ecosystems as they are absorbed for urban and industrial expansion, especially near Melbourne and Canberra to be lost completely or become extremely fragmented and persist as small remnants that are easily degraded. Many sun-moth populations occur within areas of mixed housing and industrial developments that provide challenges for management. Those commercial interests and pressures have led to energetic and politically contentious controversy over the moth's status and importance, as its possible presence on grassland provides obligations for prospective land developers to survey for the moth and if found, assess the likely impacts of their proposals in relation to the moth population size. Those needs incur both costs and delays, and bring to the fore the ambiguities over the moth's real conservation status. Discoveries of *S. plana* on many grassland sites where it seemed unlikely to occur have resulted from improved sampling protocols based on better knowledge of the moth's biology (Gibson and New 2007), especially awareness that the non-feeding moths are short-lived and that the males fly only for a few hours around the middle of the day and under rather limited weather conditions. Visual surveys at other times or in inclement weather fail to detect the moth. On any site, the flight season can extend over at least 2–3 months, but with continuously changing individuals present over this time, and each moth present for only a few days. Any 'spot sample' cannot give an estimate of the population size, and it also remains unclear whether the subterranean larvae take 1, 2 or even 3 years to develop, so that even thorough inspection of adult moths throughout a season may reveal only one of a possible three population cohorts that develop on that site. Even absence of the moth for an entire season of sampling may not mean that it is truly absent.

Degradation of native grasslands, most significantly those with abundant larval food (including Wallaby grass, *Rytidosperma* spp., previously *Austrodanthonia*) through weed invasions is a major threat to many grassland invertebrates. Some preferred native grasses (*Austrostipa*, Speargrass, and *Rytidosperma*) were maintained naturally by periodic fires and native herbivore grazing but, whilst control burning and managed stock grazing can be viable on larger 'open' areas, these may not be usable on small areas surrounded by human developments. For these, alternatives such as slashing and mowing (with machinery impacts) and herbicides may be used, with largely unknown impacts. However, the Golden sun-moth (GSM) is sometimes common on grasslands with substantial cover of the alien noxious weed, Chilean needle grass (*Nassella neesiana*), whose presence obliges attempts to eradicate it. There is strong suspicion that *Nassella* may at times be a significant food source for *Synemon* larvae (Richter et al. 2013a, b), posing the management dilemma of having to remove a declared noxious weed that has become a critical resource for a threatened herbivorous insect.

Controversies over the importance of GSM have led to policy and guidelines for development or protection of grasslands. Under the EPBC Act, a series of 'significant impact guidelines' (Commonwealth of Australia 2009) with judgements to be made on a case-by-case basis in relation to threshold values related to the area affected, the condition and isolation of that habitat patch, and the intensity of management and modification anticipated. Wider concerns over losses of native grasslands have led to several of the key ecosystems used by GSM being designated as 'Threatened communities', adding a further facet of conservation importance. The above guidelines have raised a number of conservation issues, many summarised under Victoria's 'Sub-regional Species Strategy' (DSE 2011), with commitment to rendering 80% of the highest priority habitats for GSM permanently protected. Complex issues arose over legal requirements for 'offset habitats' to compensate

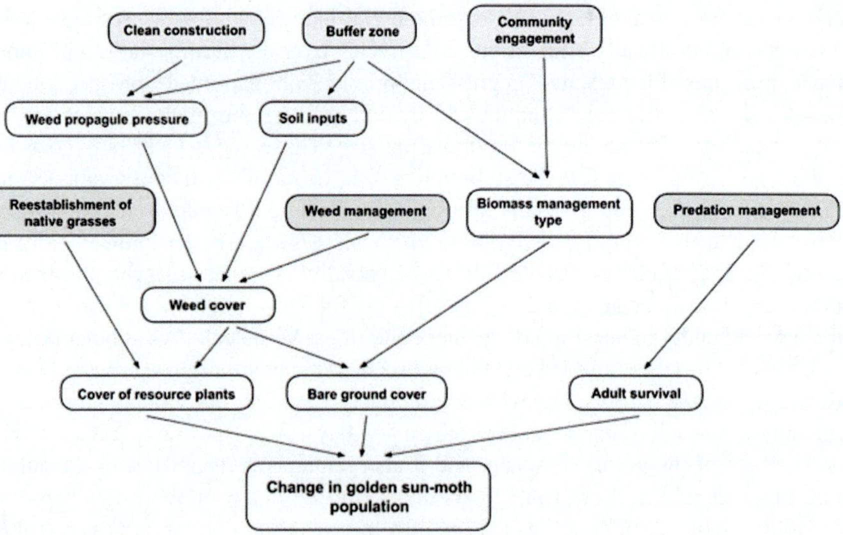

Fig. 10.1 A conceptual model indicating the web of key variables affecting change in populations of the Golden sun-moth, *Synemon plana*: darkened nodes indicate those in 'conventional management'; boxed nodes indicate 'biodiversity sensitive urban design' inputs (after Mata et al. 2016)

for those lost, and whose suitability must be assured before any disturbance of sites to be affected. These include triage of sites and deliberate sacrifice of some populations, and investigations of translocation procedures—including direct removal of sods containing early stages of GSM for implantation elsewhere. Patches of bare ground between grass tussocks are also needed to allow the poorly flying female moths to display and attract mates.

Attempts to assess the needs and practicalities for active management of GSM by eliciting 'expert knowledge' (Mata et al. 2016) led to the consensus model summarised in Fig. 10.1, as one of few such exercises undertaken for an Australian invertebrate. The inclusion of 'predation' reflected the Commonwealth recommendation that perching and nesting sites for native birds such as the Willie wagtail (*Rhipidura leucophrys*) should be minimised—such as by using open wire fences that allow passage of moths but deter birds from perching. One consequence of urbanisation is increased numbers of non-native predators on adult moths, such as the Common mynah (*Acridotheres tristis*).

Useful background references: Douglas (2004), DSE (2011), Gibson and New (2007), New (2012), Mata et al. (2016), Richter et al. (2013a, b).

Eltham Copper Butterfly (*Paralucia pyrodiscus lucida*)

Listed as Endangered under the EPBC Act, and among the first insects to be listed under Victoria's Flora and Fauna Guarantee Act, this subspecies was widely believed to have been exterminated by urban developments from the outer north-east Melbourne area from which it was described. Re-discovery on small sites scheduled for imminent housing development in Eltham led to a strenuous community-led conservation campaign focused on these small urban island sites. There, multiple threats from succession, canopy closure, weed invasions and rubbish and runoff effects compound to necessitate continuing management to sustain the butterfly, its obligatory mutualistic *Notoncus* ant and the sole larval food plant, *Bursaria spinosa*, together with monitoring of adult and larval numbers each year. Weed removal and control of succession, with careful ecological burning attempted for these purposes a consideration, are continuing management needs. The local Shire governments and Parks Victoria continue to provide strong support for maintaining and protecting the butterfly in a series of dedicated butterfly reserves in this urban region, and that also have considerable botanical significance.

P. p. lucida occurs also in a small area of western Victoria, and more widely in the Castlemaine/Bendigo region of central Victoria, where recent surveys have revealed a large population within the Greater Bendigo National Park. Conservation thereby devolves on three discrete geographical areas, with land clearing and stock grazing among the threats to some non-urban sites. Around Eltham, the butterfly is essentially conservation dependent but has become an important local emblem for invertebrate conservation.

Useful background references: Bayes et al. (2012), New (2011), Roitman et al. (2017).

Bulloak Jewel Butterfly (*Hypochrysops piceatus*) (p. 132)

Listed as Critically endangered under the Nature Conservation Act 1992, and acknowledged as one of the most extinction-prone of Australian butterflies (Geyle et al. 2021). A very localised narrow range endemic butterfly, with the main population restricted to a length of rural roadside near Leyburn, Queensland, now a dedicated nature reserve raised for the butterfly. The sole larval foodplant is Bulloak (*Allocasuarina luehmannii*), and larvae also have an obligatory association with an ant allocated tentatively to the *Anonychomyrma itinerans* species group, and now known as *Anonychomyrma inclinata*. The ant has a far wider distribution than currently known for the butterfly, suggesting that suitable breeding areas for *H. piceatus* may be found in areas not previously surveyed (Lewis and Sands 2021). The main habitat is threatened by vegetation clearing during road widening and roadside sanitation activities, and related activities on adjacent lands. Habitat protection resulted from effective liaison between state government (Department of Roads), local government and concerned entomologists.

Remnant bulloak woodland is an endangered ecological community, and an initial action plan for the butterfly urged wider exploration for further populations of *H. piceatus* and suitable habitat areas. Recent attempts have been made to use hyperspectral sensing to discriminate associated tree species and aid habitat mapping (Zainol Abdullah et al. 2014).

Useful background references: Geyle et al. (2021), Lundie-Jenkins and Payne (2000), Sands and New (2002), Zainol Abdullah et al. (2014).

Richmond Birdwing Butterfly (*Ornithoptera richmondia*) (p. 146)

Listed: Queensland: Lower Risk (Least concern), New South Wales: no concern. One of Australia's largest and most spectacular butterflies, *O. richmondia* was formerly widespread in southeast Queensland and north eastern New South Wales, but range-wide declines have resulted from forest clearance and displacement of the natural larval food plant vine by the alien South American Dutchman's pipe vine, *Aristolochia elegans*. This acts as an ecological trap, attractive to oviposition by female birdwings but toxic to young larvae. Conservation has involved a combination of removing the alien vine, planting nursery-grown *Pararistolochia praevenosa* vines (with several local nurseries recognised for producing these), protecting remaining forest remnants, and attempts at captive breeding (in part to counter inbreeding depression by releases of well-grown larvae or butterflies [bred by crosspairing individuals from different parts of the natural range] at different sites) for release. More than 500 individuals had been released by late 2020. Many schools and private landholders have participated in planting native vines and monitoring the butterfly's incidence, as a truly community-based endeavour in enriching the natural larval food supply, reducing the risks of encountering alien vines, monitoring distributions of both larvae and adult butterflies. This project, coordinated through the Richmond Birdwing Conservation Network (associated with the Wildlife Preservation Society of Queensland) is now among the longest running insect conservation campaigns in Australia (Sands and New 2013). Early years of the campaign focused largely on participation by schools, with more than 140 schools planting vines in the school grounds and monitoring for the butterfly; that exercise was coordinated through CSIRO's Double Helix organisation. Several tens of thousands of vines have now been planted as larval food for the birdwing. The butterfly has expanded its distribution into much of its historical range, and efforts continue to aid this trend by concentrating plantings of *Pararistolochia* in corridors to link existing population centres.

Useful background reference: Sands and New (2013)

Hymenoptera

Green or Golden Green Carpenter Bee (p. 156)

Formerly widespread in south eastern Australia, this spectacular bee became extinct in Victoria (seemingly from a fire in the Grampians area in 1938), and on mainland South Australia (1906). It now exists as few scattered populations in New South Wales, but with the stronghold on the western part of Kangaroo Island, South Australia. The intricate relationships between the bee, its key nesting resources and fire (p. 149) largely dictate conservation needs and demonstrate how translocations may be achieved in the future, with awareness that climate modifications and related changes in fire regimes may restrict the bees' natural capabilities to move and colonise suitable new areas.

Useful background references: Glatz et al. (2015), Hogendoorn et al.(2021).

10.4 A Wider Agenda

The species noted individually above are among those for which apparent declines and threats have stimulated conservation concern in Australia and, for most, led to actions toward conservation management. All continue to raise the public and political profile of threatened insects. They are likely to be only a very small proportion of taxa that have parallel needs but have not received, and in all probability will never receive, parallel attention. Reliance on species-level conservation for Australian insects cannot remain the major national strategy—however valuable the listing of threatened insect species may be in indicating the scale of needs and potential losses, it is logistically impossible to treat each separately or, even, to gain consensus over what the priority target taxa should be. The prominence of the cases above (and others) has flowed largely through a combination of serendipity and the zeal of individual proponents rather than from the more systematic process advocated by Taylor et al. (2018, p. 218) for selecting focal species and, as Taylor et al. emphasised, the full array of listed insect species in Australia do not adequately reflect the variety of either taxa or ecosystems vulnerable to anthropogenic influences. Their plea for more systematic selection of focal taxa to represent all bioregions would do much to restore balance—but raises formidable practical problems in pursuing practical evaluation and conservation. Increasing the capability needed to adopt a wider practical perspective for insect conservation is integral to any such suggestion. The increased ecological and geographical representation recommended by Taylor et al. would markedly increase the number of listed species, but also increase the already formidable logistic gap between demonstrated need and capability to address this. Trends in the composition of current insect listing in Australia generally mirror that in Europe, where strong taxonomic bias has led to many orders being wholly or almost wholly ignored, and the tiny proportion of species listed or evaluated also unlikely to fully represent conservation needs of even closely related species. The few insect species listed in Europe (123 of 105,000, p. 80: Leandro et al. 2017) represent only seven orders, and taxonomic bias within these can also be strong: toward butterflies in the Lepidoptera, for example, whilst few moths were listed. Surveys across several categories of insect features in Europe to assess what characteristics may be associated with 'listing', in addition to being threatened, found that protected species tended to be larger, better known, more widespread and more multicoloured than a random set of 123 unlisted species. Ecological biases were also evident, with nectar-feeders and dead wood species proportionately over-represented. Leandro et al. (2017) examined the European (Bern Convention) criteria for listing species, which were founded in the credibility from species' vulnerability, practicality (easy to recognise and identify) and public appeal (size and colour as features fostering public support). Much the same trend has occurred in Australia and, equally, has been accompanied by neglect of most insect taxa that do not garner wider attention, through being poorly known, small and 'dull' coloured—and simply having no strong advocate for their wellbeing.

The present largely ad hoc approach to selecting focal species has indeed provoked much advance in insect conservation interests—but many deserving insects are likely to evade attention through factors such as lack of clear taxonomic recognition, the vulnerability of restricted sites whose denizens are largely unknown, occurring in remote areas where their changed status is unobserved, or simply being members of taxonomic groups that do not gain popular attention. And, fundamentally, the number of deserving putatively threatened candidate species that could be nominated validly for listing could not be given adequate attention from the resources now available or likely to become so in the forseeable future. Most of the rather few insect species currently listed do not receive sufficient attention, in some cases, none. Any more systematic approach to selecting species for listing can retain credibility only by continued focus on 'popular' or recognised flagship groups for which at least framework documentation and sympathy exists. At present, those are the only species with any prospect of attracting conservation funding, this essentially preventing or substantially thwarting further observations toward conservation status evaluation. Whilst 'the species approach' remains the most tangible focus for gaining citizen support and interest, the candidates must be accepted as worthy, in need of conservation and able to have their chances of decline or loss reduced by activities that are practicable and be monitored effectively.

That principle effectively excludes the vast majority of insect species from individual consideration. However vulnerable, scarce or restricted some may seem, it is unrealistic to project that the numerous poorly-documented members of most insect groups will ever be singled out for individual attention, despite advocacy for their importance and susceptibility from concerned specialists (in line with the understandable trend for such specialists to promote interests in 'their' group'). These taxa are the 'meek inheritors', with features likely to accord them political anonymity in conservation (New 2000): thus, they may not bite or otherwise harm people, have no tangible 'pest' status or proven ecological role, are commonly small and obscure (perhaps needing specialised techniques or experience to capture and study them), hard to identify, and have no attraction or commercial value to collectors. Further, their biology is usually unknown in any detail and their roles in food webs and ecological processes equally obscure. As components of large and often uncharacterised biodiversity, nominating such species individually for conservation may lead to loss of credibility when attempting to justify that action. Alternative conservation strategies are clearly necessary, and the broader approaches of conserving them under some more credible 'umbrella' take precedence. A more fundamentally ecological perspective, in which key environments become a major focus, has considerable potential and represents a paradigm shift (Leandro et al. 2017) in expanding from species-focus to wider scales of the environments where insects occur. Thus, the focus on species, as the 'fine-filter' level of conservation leads to the 'meso-filter' level (of focus on local features, such as ponds, patches of native grassland or forest, small hills and other topographical or ecological units) often relevant to defining the habitats of the focal species and themselves defined by those needs, and across which much relevant experience may have accumulated. The broader diversity of those areas is the main concern. Finally, the 'coarse-filter' level of landscape conservation is clearly a primary aim to embrace insect diversity, for which basic conservation principles have been suggested repeatedly. Part of Samways' (2020) 'Future-proofing' (see preface, p. v) noted the need to set clear goals for conservation actions and continually seeking strategic ways for advance. He noted a series of principles for landscape conservation as (1) maintaining protected areas; (2) maintaining high-quality heterogeneity in landscapes; (3) softening habitat edges to reduce contrast and abrupt changes and produce more gradual boundaries between habitats; (4) softening transformed areas such as agricultural and urban developments by preservation or introduction of green spaces and native plants, such as on field margins, hedgerows, roadsides and urban parklands; (5) improving functional connectivity across the landscape; and (6) simulating natural disturbances through analogue activities such as burning, mowing or slashing, selective tree-felling and others tailored as management tools. Such disturbances in nature have moulded the persistence, interactions and resource needs of insects over evolutionary time and contribute to maintaining larger and well-connected habitat patches within landscapes. 'Traditional' low intensity management of agricultural and forest ecosystems is recognised as key to insect conservation in parts of Europe, and those lessons have wide relevance elsewhere where long-term low intensity management has been practised. Collectively, the above principles comprise a 'synthetic landscape conservation approach' that can accommodate the needs of numerous insects. Sustaining ecosystem function is a key driver of this wider perspective—and Leandro et al. (2017) emphasised that this is very different from the widely suggested 'ecological services' that are essentially considered in terms of anthropogenic interests rather than wider ecological integrity. However, Leandro et al. noted three new 'ecologically based' criteria that could augment the Bern Convention spectrum to increase functional representation of insects on protected species lists, as (1) umbrella features, species whose presence indicates greater diversity; (2) keystone features, species with critical ecological significance in the ecosystem they inhabit; and (3) community indicators, species whose presence indicates 'healthy' communities. These collectively suggest a more inclusive approach to insect conservation, as discussed below.

The species cases considered above almost wholly represent taxa from temperate regions, in areas where there are sufficient interested people to evaluate and pursue conservation need. This close attention to individual species, the backbone of insect conservation developed in northern temperate countries, differs from the primary need over most tropical regions, in which such close focus is largely impossible. Even for butterflies, the major conservation strategy in the tropics has, rather, been one of habitat protection (Bonebrake et al. 2010). The detailed studies of species and population trends that form the basis for most temperate-region species conservation programmes are scarce among tropical insects, and seem unlikely to proliferate in Australia, so that continued focus on habitat protection must remain a priority. However, that protection necessarily acknowledges that tropical environments are complex, the richness and abundance of their resident insects is high and largely unknown Lewis and Basset 2007), that the multiple interactions in which those insects participate are almost wholly undocumented (but many likely to be specific or mutualistic), and that the influences of invasive species, climate change and other factors on insect populations and communities are only tenuously predictable.

As elsewhere in the world, opportunities for considerably increasing the extent of the declared protected areas are limited progressively by other demands and by wide degradation of landscapes and waterbodies on areas potentially available. Much of the future of insect conservation will depend on modifications to agricultural, forestry and urban areas, in collaboration with the varied stakeholders and private interests. Perhaps especially in urban areas, conflicting demands for land and escalating prices of land purchase increasingly create barriers to conservation. However, habitat 'networks' (interconnected individual units across a local or wider landscape) are a key aspect of conserving beta diversity, recognising that individual patches differ in their species complements and diversity but that all contribute to sustaining a regional biota. Urban vegetation remnants and ponds both reflect that individual patches may gain attention and conservation priority because of high insect richness, but considering them more widely as networks provides additional benefits. Particular criteria may be applied for conservation: ponds in the United Kingdom are required to support several rare species and a total of >50 aquatic invertebrate species, as well as meeting other demands (Hill et al. 2021).

Two main components found beta diversity: (1) richness differences across patches (with conservation importance generally directed to sites with greatest richness; although low richness on sites may comprise a subset of the species on higher richness sites, they may gain significance by harbouring rare species not widely represented elsewhere); and (2) turnover, or species replacement, reflecting the gains and losses of species among sites as conditions and opportunities change. The two components have rather different implications for conservation. If the low richness of a patch truly represents a subset of the regional high richness species, effective conservation of the high richness patches may suffice. However, if species replacement is high, with the patches differing more in species representation, a range of sites may be needed for adequate regional conservation.

Collectively, landscape-scale conservation supports the aim of highest overall protection of biodiversity. Hill et al. (2021) advocated pond networks as 'ideal systems' to assess conservation strategies because they are abundant and clearly bounded within the landscape, and also implied that pond conservation would be most effective at the landscape level. This may become difficult in practice, especially in urban areas, so that the contributions of individual sites to overall landscape conservation help to set priorities. In their study across 132 ponds in five urban areas of England, 70–90% of the regional species pool were protected if ecologically unique and high taxonomic richness (>50 aquatic species) were considered together, but 54–94% when only sites with high richness were included (Hill et al. 2021). The clear values of ecologically unique ponds (reflecting their contribution to beta diversity) in conjunction with high richness sites led Hill et al. to list four positive advantages of this approach, as (1) supporting landscape-level conservation as those sites with highest contributions to beta diversity are protected; (2) continuing to ensure that sites with highest biodiversity are protected; (3) ensuring that a significant proportion of the regional species pool is protected; and (4) expanding the network of protected aquatic habitats in urban areas. Attention to 'ecological uniqueness' may also lead to conservation of ponds in areas where none are at present protected.

The principles of 'biodiversity sensitive urban design' (Garrard et al. 2017) recognise the need and urgency for conservation across the spectrum of urban environments within a milieu in which conservation attention is constantly competing with development interests, through documenting ecological values and biodiversity characteristics and seeking to harmonise these with such competing pressures and to overcome the view expressed by Hall et al. (2017) that 'Cities are primary viewed in terms of their political value (where the votes are) rather than for their ecological value'. Urban design incorporating sound ecological principles embraces all habitats of native insects, with the above pointers for ponds extending to terrestrial vegetation remnants and, more widely, vegetated open spaces. Values to insects of natural vegetation remnants, in particular, are accepted widely. Urban expansion in Australia is largely into relatively natural periurban areas, such as native grasslands around Melbourne and Canberra (that can support diversity and insects of conservation significance), although a few older enclaves have been set aside for conservation within urban areas. All these remnants can contribute to an effective habitat network, and the values of remnant native vegetation in towns and cities may be far higher for insects than more anthropogenic open spaces such as parklands and many gardens (New 2015). Recent surveys in Melbourne compared the incidence of some key insect pollinator taxa in a natural vegetation site (reserved in the early 1960s) and more unnatural vegetation sites in catches from coloured pan traps (Shrestha et al. 2021). Bees considerably outnumbered hoverflies (Syrphidae), but both were more abundant in the remnant area, and native bees were much more abundant in the native vegetation (Fig. 10.2).

However, practical attention to the values of some taxa strengthens need to understand them better and to accelerate or improve taxonomic and biological knowledge to better evaluate richness and distributions. Monitoring environmental quality and the impacts of disturbance often involves using insects as 'indicators' (or some form of surrogate) to assess environment condition and responses to changes. For freshwater ecosystems, for example, changed conditions within complex communities have been estimated commonly by abundance and richness of selected aquatic insect groups. Characterisation of standard samples of larvae of mayflies (Ephemeroptera), stoneflies (Plecoptera) and caddisflies (Trichoptera), as the 'EPT Index' (p. 36), is taken commonly as a measure of aquatic environment quality and monitored to evaluate disturbance impacts. A fourth aquatic order, Odonata—as a far better-documented group for non-specialist access—augments these in some studies. The 'Dragonfly Biotic Index' (DBI) (Samways and Simaika 2016) is an index of freshwater condition, based largely on the conspicuous adult stage—and the generally straightforward recognition of most odonates to species level gives it considerable sensitivity. Use of the EPT Index in Australia has been widespread (New 2021) and representation of these selected taxa infers condition of wider aquatic biodiversity. The DBI has not yet been applied thoroughly in

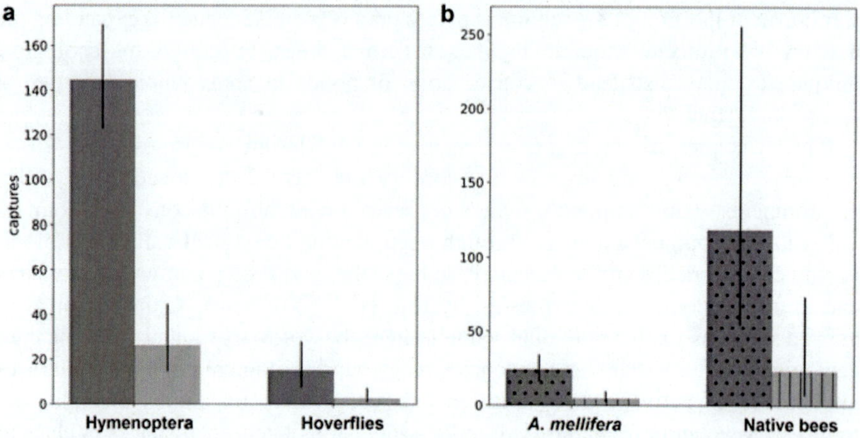

Fig. 10.2 Comparison of numbers of Hymenoptera and hoverflies (Diptera: Syrphidae) captured in area with remnant native vegetation (left hand bars) and an urban residential location (right hand bars): (**a**) mean numbers captured at both sites; (**b**) mean number of *Apis mellifera* and native bees at both sites (modified after Shrestha et al. 2021, used under Creative Commons Attribution [CCBY] licence: https://creativecommons.org/licences/by/4.0/)

Australia, but its use in South Africa demonstrates the constructive potential for understanding it offers to managers: the comprehensive field guide for species dragonfly identification (Theischinger and Hawking 2021) facilitates this adoption in Australia, using protocols based on those described by Samways and Simaika. As with any such exercise in monitoring or evaluation, the question to which answer is sought must be articulated very clearly, and the general principle of recording all species found at the site(s) assessed followed as background for biodiversity recording and any statistical analysis. The DBI is useful in ranking habitat quality and allocating a 'Habitat Condition Score' across series of sites.

Indicator groups of insects have been suggested or claimed for many ecological disturbances, and are strong incentive for focus of future taxonomic studies and evaluation as components of ecological monitoring. Although not specific conservation targets, local declines of species used for such monitoring can herald much wider effects. The EPT-approach is based on the notion that responses of the constituent species to environmental condition and change, embracing ecological, physiological and behavioural responses, will consistently reflect the responses of a larger taxon set that is not itself measured, in terms of species richness and assemblage or community composition. It has been claimed to have considerable value in freshwater faunistic studies.

The wider landscape-level approach includes protection of notable communities and promoting connectivity by enhancing intervening areas and reducing the threats they present, to build on the conservation provided by formal protected areas. The twin intergrading principles of 'land-sparing' and 'land-sharing', applied most commonly to increase conservation values of agricultural landscapes, exemplify much of the ecological understanding underpinning insect conservation. Land-sparing refers to the distinct designation of land for conservation, so sacrificed from production areas. Those lands may include 'set-aside', field margins and headlands, vegetation strips, potential corridors and others, and some may become formally protected. Land-sharing seeks to accommodate conservation measures within the agricultural activities, with measures drawing on conservation biological control and reduced management intensity—such as reduced pesticide use and weed control, organic production, removal of alien invasive species, and other changes that foster more natural conditions amongst which native insects can survive and generally increase heterogeneity across the areas treated. Together, these principles contribute to an agroecological strategy that includes the various practical needs involved, reduces some key threats to natural biota and enhances/restores some critical resources for native insects—all in the context of increasing heterogeneity and fostering greater connectivity. Insect conservation in agricultural areas has channeled much effort into enhancing ecological services for crop production and crop protection, and has involved development of strategies that provide additional resources—flowers for pollinators, native vegetation for herbivores, prey and hosts for natural enemies of crop pests and increasing their access to adjacent crop environments—with insects benefitting widely from measures such as preserving, enhancing or creating features such as hedgerows, uncultivated field margins, and additional native flora. Structures such as 'beetle banks', raised rows of natural grassland and associated native vegetation extending from field margins into the crop, facilitate movements of natural enemies into those crops.

Diversity of flora in field surrounds can significantly improve crop pollination levels, with flower variety and distance to crops both influential—for planted flower strips, pollination effectiveness declined exponentially with distance (Albrecht et al. 2020). The trends were illustrated in England, where a six-year survey of crop yields and the abundance of pollinators and natural enemies differed across three treatments—unmodified ('business as usual', control) and with 3% and 8% of cropland at field edges taken out of production and sown with wildflowers to increase quality of the areas for insects (Pywell et al. 2015). Increased crop yields over the period compensated for loss of growing area, and the abundance of both pollinators (bees) and predators (carabid beetles) of crop pests increased substantially in the more extreme treatment (Fig. 10.3). Aims of making near-field environments 'bee-friendly' have emphasised the values of such multifunctional areas, whose roles in enhancing insect diversity are recognised widely as both sanctuaries/refuges in highly altered environments and enabling local insects to persist and move through the landscape. Both in agricultural and urban areas, increased variety of native flora, opportunities for bee-nesting or other shelters, resources such as retained dead wood,

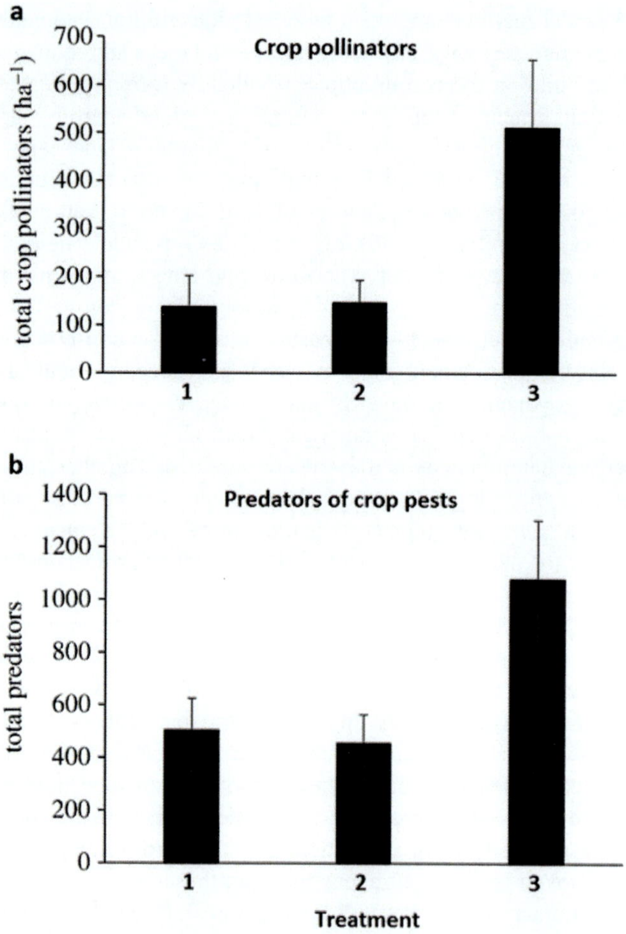

Fig. 10.3 The mean (± s.e.) abundance of (**a**) pollinators (*Apis mellifera* and *Bombus* spp.) of oilseed rape and field beans per hectare, 2007–2010; and (**b**) predatory carabid beetles in 2008. Treatments (see text) are 1, business as usual; 2, 3% of crop area and 3, 8% of crop area rescued (after Pywell et al. 2015)

reductions in pesticide use and pollutant run-off, increasing buffer margins and generally increasing the 'mosaic nature' of the area can provide enormous benefits.

Vegetated strips have other roles focused less directly on agriculture, such as water protection by mitigating soil erosion and reducing leaching of agricultural and industrial chemicals that might at times reduce potential for biodiversity (Haddaway et al. 2016), and may need management consideration. Preventing nutrients and sediments from reaching waterways may become the most significant conservation measure and aim of riparian or other vegetation as buffer plantings. High quantities of nutrients and pesticides can severely affect insects in freshwater bodies and may be transferred by flow far downstream from their entry point (New 2021). However, any such plantings can also harbour insects and contribute to landscape connectivity. Likewise, urban plantings of native vegetation are advocated frequently, and replacement of existing woody vegetation and mown lawns on nature strips by wildflower combinations can foster significantly increased arthropod abundance (Mody et al. 2020, for Germany) and also reduce maintenance costs.

Studies continue to proliferate and to endorse the conservation values to insects through such modifications. They have also confirmed that land-sharing and land-sparing are not mutually exclusive, as both contribute to the multiple functions needed for conservation-friendly agrolandscapes. Land-sparing is vital only for those species (usually ecological specialists and so potentially vulnerable) that are highly incompatible with agricultural modifications. Otherwise, high connectivity is a primary need so that (1) ecosystem service providers (notably pollinators and natural enemies of pests) can spill from protected or managed areas to production areas and, with others, can be rescued from losses associated with isolation; (2) immigration can counter possible extinctions in spared areas; and (3) conserving community diversity assures that ecosystem services continue as conditions change (Grass et al. 2019). Insects and others must be able to 'track' environmental changes and colonise further suitable patches as conditions change. This realisation may necessitate changes in the predominant practice in Australia where, historically, conservation has been focused largely on protected areas, whilst extensive agricultural areas have been devoted wholly to production—so that land-sparing has been the main approach—notwithstanding subsequent lack of formal conservation activities and management. Very broadly, a landscape approach to conservation, aiming to harmonise and reconcile the various competing social, economic and environmental objectives of the various stakeholders (Chatterton et al. 2016) necessitates a long-term commitment of cooperation, leadership, interests, and clear goals to protect biodiversity and the wide range of ecosystem processes. Energetic short-term interests for economic reward may be a major hindrance to any such enduring aim (Dudley and Alexander 2017).

Most insects remain as passengers in such exercises, but increasing awareness that many species occur as metapopulations has emphasised the values of this wider approach to their conservation. The acceptance of a considerable variety of threatened ecological communities designated under Australian legislations provides opportunity to further protect sensitive habitats of many specialised insect taxa, but—other than for notable individual species that can add weight to cases for protection—the detailed features of insects within those communities are usually incompletely known. A plea for inventories of insects in Australia's national parks (Sands and New 2003, p. 126) extends equally to other protected areas, and was prompted in part by a statement by Yen and Butcher (1997) that 'Very few reserves in Australia, if any, have had extensive invertebrate inventory to simply determine what is being conserved, let alone long-term monitoring to test the effectiveness of the reserve system'. Most of the sporadic surveys that have been accomplished have been taxonomically restricted, single occasion unreplicated events undertaken over short periods and not followed up by additional exercises, and so are 'snapshots' of a continuously dynamic scenario. Many have also been initiated by individual scientists or other enthusiasts rather than as an initiative or recognised routine 'duty' of management staff. Likewise, they focus on particular groups of interest to the proposer(s) rather than more comprehensive coverage. Even for butterflies, and reflecting the sometimes dauntingly complex system of applying for permits to collect and study in national parks, Sands and New commented that 'we know far more of the distribution of most butterfly species outside national parks than within those sanctuaries'. In Australia's national 'Butterfly Action Plan', Sands and New (2002) noted need for additional surveys simply to clarify the conservation status of the 105 threatened taxa by determining their incidence in protected areas. This principle is by no means novel in Australia but, for invertebrates, has been slow to develop from ideal to practice. Many experienced entomologists have a 'reasonable idea' of what taxa of their interest group may be present in a given protected area, but detailed surveys, with the data made widely available to managers, would provide a sound template for insect enumeration and conservation evaluation, and a basis for sensible monitoring of changes. Ideally, but impractically, those surveys would encompass all orders of insects as a basis for examination for specialists to evaluate the significance of the fauna, the presence of especially notable species or richness, and the vulnerability of the different groups in that particular environment. In practice, this level of penetration is unlikely to occur: both specialists and financial support are in very short supply! There is thus need for some consensus over which insect groups would most repay such investigation through facilitating and optimising conservation management, not least so that the limited resources available can be channeled for greatest returns and conservation benefit.

Many entomologists would nominate or advocate the relevance of 'their' favourite insect group for wider ecological interpretation or surrogacy values, based on personal experience and predilections, to reflect wider diversity or community condition and change. However, proving this amongst the immense terrestrial insect faunas of Australia, with their numerous poorly known (or unknown!) species that are largely unrecognisable and unidentifiable other than by their champion specialists, creates problems of credibility and proof. Whilst some insect taxa, from species to far larger groups, may be especially characteristic of given habitats (many of them defined as vegetation associations), any surrogacy or indicator values are mostly poorly defined—or, even, just 'wishful thinking'—amongst such incompletely understood arrays of taxa. Use of ecological arrays not necessarily founded in individual taxonomic entities may also have a place in assessing diversity, and acknowledges the limitations of relying only on one or few purported indicator species.

The broader approach of 'predictor sets' (initiated by Kitching 1993 and discussed in detail by Kitching and Ashton 2013) seeks to incorporate ecological variety in selecting a limited range of insect species that collectively capture a varied array of biological responses to environmental changes, and that could become the main focus for monitoring in the particular context of interest. Ideally, the taxon set should contain, as far as possible, a range of functional groups that capture the biological variety of the assemblage they represent, and the members should be sampled readily and be identifiable unambiguously. The choice of taxa of the predictor set should be predefined in some quantifiable way, and the set must comprise taxa (of any level) whose statistical distribution within the wider assemblage has at least the same relationships with the focal environmental variable as does the entire assemblage (Kitching and Ashton 2013). One practical aspect is that a predictor set can be defined only from the basis of comprehensive baseline information—so that comprehensive survey of the target assemblages from which they are to be designated is necessary, even to reveal the higher taxa (such as families or subfamilies) from which predictors may be drawn. Likewise, a given predictor set may be valuable only for a limited range of changes or inter-site environmental differences, so that extrapolations to other contexts must be cautious.

A critical need for any predictor or indicator suite of taxa is consistent and accurate recognition of the entities involved, whether or not they can be identified formally. Without that certainty, the exercise fails. In the interests of facilitating recognition, it may be necessary to 'edit' the array of taxa used—thus, one of the predictor sets of moths derived to assess elevational differences in assemblages in Queensland was reduced from an initial set of 28 species filling the basic requirements of adequate abundance and restriction to particular environments, to 18 that could be identified reliably but then to only nine that could be recognised clearly by non-specialist sorters without delay (Kitching and Ashton 2013).

Lack of fundamental biological information is perhaps the most universal hindrance to pursuing any form of insect conservation above the species level in Australia. Even for butterflies, as the most fully investigated insect group, a series of 'recommendations' summarising the major needs for advancing conservation assessments and practice (Sands and New 2002) (Table 10.1) displayed serious gaps in documentation, and these are far larger for most other insect groups—including those recommended for monitoring and inventory studies. Parallel sentiments continue, seeking to address the impediments discussed earlier (Chap. 2) but also expressing the frustrations currently inherent in establishing the most rational and well-focused agenda for Australia's insect conservation. In the more specific context of appraising invertebrate responses to fire, Saunders et al. (2021) noted the 'insufficient evidence about invertebrate diversity and ecology' as limiting capacity to form evidence-based decisions over post-fire land restoration and funding

Table 10.1 The seven recommendations of priority needs for improving understanding of the conservation status and needs of Australian butterflies (Sands and New 2002; New and Sands 2004)

1. Need for additional surveys
2. Need for additional biological information
3. Need for habitat security
4. Need for habitat restoration
5. Need for threat evaluation and management
6. Need for effective planning and execution
7. Need to recognise purposes and roles of listing species, and subsequent actions

priorities. Similar comments could be made in a variety of other contexts of post-disturbance ecosystems spanning terrestrial and inland water environments. Saunders et al.'s priorities included research and policy/decision-making needs and, although framed primarily for fire, apply to many other threat appraisals in which greater understanding of preventative and restorative management on insect diversity and vulnerability is currently lacking.

The diversity of Australian insects, that the biology and distribution of many of them is inadequately known, the taxonomic impediment, and the generally low levels of political and public acceptance of the importance of insects (and other invertebrates) to Australia's ecosystems collectively comprise formidable hindrances to advancing insect conservation and truly assessing the needs for it to be accelerated (New and Yen 2013). However, increasing recognition of these and related deterrents continues to stimulate concerns and advances that expand both scientific knowledge and the involvement of people in all walks of life. Some of the major species' case histories summarised earlier demonstrate the involvement of strong networks and groups of concerned citizens. Perhaps the greatest logistic problem—beyond the obvious lack of funding and the low number of conservation managers experienced in entomology and with the time to commit to insect conservation—is simply the effective transfer of available information from basic documentation to field contexts where it can be applied by agency and other personnel.

A call for each State and Territory conservation agency to employ at least one full-time invertebrate expert, initially as a recommendation by Yen and Butcher (1997) and reiterated several times since then, has not eventuated, despite some increased interest. Suggested roles of such persons listed by Yen and Butcher (and largely endorsed by contributors to a later workshop: Cassis 1999), all within the mandate of promoting invertebrate conservation and coordinating relevant interests, included (1) formulating policy on invertebrates, including relevant legislation; (2) preparing a State or Territory overview of invertebrate conservation needs, collating all available information; (3) establishing and maintaining a register and data base of all relevant invertebrate research projects, literature (both published and consultant reports), threatened taxa, communities and habitats and sites of invertebrate significance; and (4) assisting in obtaining funds for invertebrate research. Reflecting the paramount popularity of butterflies as conservation targets in Australia, liaison between lepidopterists and agencies could be facilitated strongly through a designated 'invertebrate person' in each agency (Sands and New 2002). However, the contrasting trend of 'downsizing' the core personnel of many management agencies pushes insects still further into the background rather than promoting their interests through dedicated attention and support. Wider communication between scientists and managers is urgent, not least to promote trials and acceptance of recent scientific advances and discovery. The clear elucidation of predictor sets in one of the projects (funded by a management agency) discussed by Kitching and Ashton (2013), for example, was not taken up by that agency and three speculative reasons were noted for this: (1) despite the recognised needs for long-term monitoring, the funding and will to do this was not there; (2) diverting attention to non-charismatic and unfamiliar taxa such as arthropods was perhaps not competitive with accommodating the needs of high profile taxa such as tree kangaroos, crocodiles and cassowaries; and (3) the particular 'champion' for arthropod monitoring had transferred to another position within the agency. The latter is a widespread limitation as staff are promoted or removed and replaced (if at all) by people without equivalent interests or experience. Such inconsistent or saltatorial support and priority renders continuity of many worthwhile exercises highly uncertain. Turnover of staff, many of whom are likely to have only limited experience with invertebrate conservation, suggests that continuing focus on those species with stronger documentation, conservation record and experience in management will continue to have a predominant role, and that additional taxa and environments lag. The increases in focal species coverage recommended by Taylor et al. (2018), incorporating both threatened and iconic species sufficient to represent all major categories and regions of insect habitats in Australia is highly desirable. The model presented as a foundation strategy (Fig. 10.4) shows a clear way toward progress. This approach acknowledges that candidate species are selected, in addition to primary need, on geographical and ecological characters and across a variety of habitats—and that the limited array of currently listed species is inadequate for any realistic faunal representation. Increasing the ecological coverage parallels needs to augment the initial insect listings under the Bern Convention in Europe, for which many key ecosystems and taxa were markedly under-represented and neglected but difficulties of expanding that coverage complex (Haslett 2012). Earlier, Haslett (1997) noted invertebrates of cave ecosystems, saproxylic environments and running waters as under-represented. The listing predominance of butterflies in Australia can mask needs for other taxa and biotopes other than the limited range of terrestrial systems those species frequent—but also displays the variety of needs within this single group and the range and scale of conservation activities needed and that can be projected for 'insects as a whole' (New 2009).

The concerns for insect conservation in Australia, evident increasingly over the last few decades, encompass a great variety of themes and stakeholders with very different priorities and interests. Integrating these toward the common purpose of assuring a safer future for insects and their multitude of needs in the face of escalating human pressures and the uncertain overlay of climate change impacts still has far to go. Continuing efforts to enumerate and diagnose insect diversity, its distribution and needs in the face of current and future threats contribute to a firmer template for action drawing also from the contexts and cases noted in earlier chapters and contributing to more fundamental knowledge and awareness of insects, and development of more reliable and tested approaches to their conservation. Gradual lessening of the various 'impediments' discussed earlier, including acceptance of the unique nature of Australia's insect fauna within both government and citizenry and linked with energetic education measures to enthuse young people, is a vital and welcome trend. However urgent it may be, building the scientific capability to achieve these goals is uncertain. Even for Europe, Haslett (2008) could comment on the decline of relevant training in educational institutions as emphases shifted away from traditional study of classification and identification of life forms at the very time that governments were becoming aware of the needs to protect

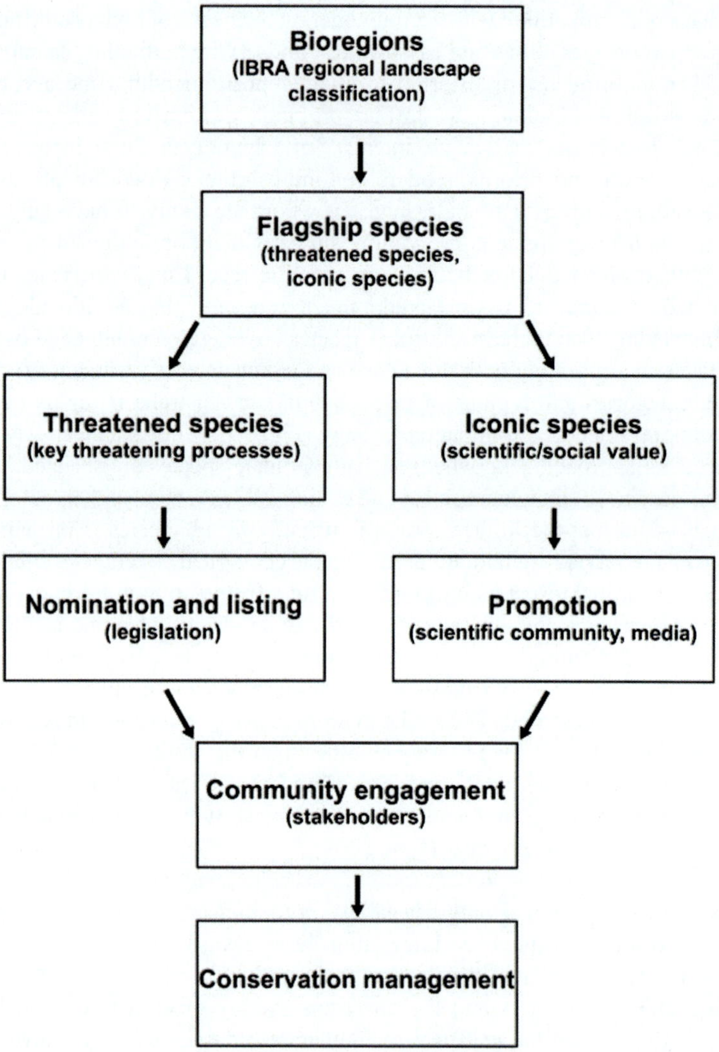

Fig. 10.4 A strategic framework for conservation management of Australian insects, based on individual species (after Taylor et al. 2018)

biodiversity. The needs for specialist knowledge to identify, recognise and describe the species of most insect groups are now greater than ever before. Haslett's comment, again for Europe, that 'threats to taxonomist extinction are severe and increasing' and inference that this, in turn, is a severe threat to the entire invertebrate conservation effort leaves little room for complacency. Despite valiant and ambitious planning (AAA/RSTA 2018), most insect groups in Australia will be poorly documented passengers in conservation activities for the forseeable future.

The strategy advanced and discussed by Taylor et al. (2018) focused on optimal selection of species to achieve greater representation of suitable insect environments (bioregions) throughout Australia, but could not overcome the utopian need for

Table 10.2 The objectives of the European strategy for the conservation of invertebrates (Haslett 1997)

1. Raise awareness and alter human attitudes and behaviour towards the importance of conserving invertebrate animals
2. Promote integrated management of landscape mosaics at the relevant scales to be sustainable for invertebrates
3. Strengthen European to national/sub-national invertebrate conservation policy and action
4. Identify and prioritise key actions to be implemented at different political and geographical scales
5. Promote accessibility and efficient flow and exchange of information on invertebrates within and between the scientific and public domains
6. Promote inclusion of a fully representative variety of invertebrate species in conservation and environmental management decisions, including integration of invertebrate conservation into existing and future conservation strategies involving other groups of organisms
7. Build scientific and technical capacity for the conservation of invertebrates and identify areas of urgent further research

adequate political and public support for integrating and undertaking such a venture, other than by increased numbers of concerned 'nominators' of species, and central coordination of activities. A wider strategy could be founded in the objectives of the pioneering European Strategy (Haslett 2008) (Table 10.2), which collectively express many of the issues foreshadowed by earlier commentators and are implicit in the earlier discussion. The Taylor et al. strategy addresses mainly Objective 6 of this table, as reasonably tangible and embracing much of the background ecological variety present and demonstrating the variety of concerns for insects themselves. Nevertheless, the European strategy objectives raise many themes of wide concern in Australia, and their relevance and practicality deserve serious evaluation as insect conservation based on increasingly complete evidence of needs is developed. The key actions for each objective suggested by Haslett (2008) give many pointers of potential value elsewhere. The international cooperations needed for unified actions across Europe are paralleled by needed consistency across the State and Territory divisions of Australia, with the Commonwealth providing national leadership. Greater recognition of insects within the EPBC Act and greater uniformity of evaluations across more local jurisdictions, as an anticipated consequence of the Taylor et al. (2018) recommendations, will be an important step toward this wider perspective.

The increasing concerns and interest in insect conservation in Australia, based on wider community participation and understanding, are encouraging, but considerable challenges remain. The welcome trends noted above, essentially signalling the urgency and needs to document the resident fauna more completely and embed insect conservation (together with sustaining the varied habitats on which insects depend) more firmly in consistent policy, go hand-in-hand with threat mitigation and prevention. Realisation that all these (and related activities) must proceed without complete documentation of both the biology and diversity of insect targets and the outcomes of largely untried management approaches does not obviate the urgent need for actions based on the best knowledge and experience available and documentation of those actions and their underlying rationales for later reference. Without those actions, many of Australia's localised and endemic insects must almost inevitably decline over the next few decades, and their eventual losses diminish the country's unique biological heritage.

References

AAS/RSTA (Australian Academy of Science/ Royal Society Te Aparangi) (2018) Discovering biodiversity: a decadal plan for taxonomy and biosystematics in Australia and New Zealand, 2018-2027. Canberra, Wellington

Ahern LD, Tsyrlin E, Myers R (2003) Mount Donna Buang wingless stonefly *Riekoperla darlingtoni*. Flora and Fauna Guarantee Act 1988, Action statement no 125. Department of Sustainability and Environment, Melbourne

Albrecht M, Kleijn D, Williams NM, Tschumi M, Blaauw BR et al (2020) The effectiveness of flower strips and hedgerows on pest control, pollination services and crop yield: a quantitative synthesis. Ecol Lett 23:1488–1498

Baird IRC (2012) The wetland habitats, biogeography and population dynamics of *Petalura gigantea* (Odonata: Petaluridae) in the Blue Mountains of New South Wales. PhD Thesis, University of Western Sydney. http://handle.uws.edu.au:8081/1959.7/509925

Baird IRC, Burgin S (2016) Conservation of a groundwater-dependent mire-dwelling dragonfly: implications of multiple threatening processes. J Insect Conserv 20:165–178

Bayes E, Douglas F, Whitfield J, Van Praagh BD, Field RP et al (2012) Surveys for the Eltham copper butterfly *Paralucia pyrodiscus lucida* Crosby (Lycaenidae) in Victoria in late 2011. Vict Nat 129:114–118

Bonebrake TC, Ponisio LC, Boggs CL, Ehrlich PR (2010) More than just indicators: a review of tropical butterfly ecology and conservation. Biol Conserv 143:1831–1841

Cassis G (1999) Workshop: putting invertebrates on the agenda: opportunities for the future. In: Ponder W, Lunney D (eds) The other 99%. The conservation and biodiversity of invertebrates. Royal Zoological Society of New South Wales, Mosman, pp 437–445

Chatterton P, Ledecq T, Dudley N (eds) (2016) WWF landscape elements: steps to achieving integrated landscape management. Worldwide Fund for Nature, Vienna

Cleave R, Tulloch C (2015) Phasmid. Saving the Lord Howe Island stick insect. CSIRO Publishing, Clayton South

Commonwealth of Australia (2009) Significant impact guidelines for the critically endangered golden sun moth (*Synemon plana*). EPBC Act policy statement 3.12. Department of the Environment, Water, Heritage and the Arts, Canberra

Cordero-Rivera A (2013) Behaviour and ecology of *Hemiphlebia mirabilis* (Odonata: Hemiphlebiidae). Report on work undertaken. Department of Environment and Primary Industries, Melbourne, VIC

Cordero-Rivera A (2015) Demographics and adult activity of *Hemiphlebia mirabilis*: a short-lived species with a huge population size (Odonata: Hemiphlebiidae). Insect Conserv Divers 9:108–117

Crowther D (2011) The ancient greenling: synthesis of new information to improve conservation outcomes. Arthur Rylah Institute for Environmental Research, Heidelberg

Douglas F (2004) A dedicated reserve for conservation of two species of *Synemon* (Lepidoptera: Castniidae) in Australia. J Insect Conserv 8:221–228

DSE (Department of Sustainability and Environment) (2011) Sub-regional species strategy for the Golden sun moth – draft for public consultation. Department of Sustainability and Environment, Melbourne

Dudley N, Alexander S (2017) Agriculture and biodiversity: a review. Biodiversity 18:45–49

Garrard GE, Williams NSG, Mata L, Thomas J, Bekessy SA (2017) Biodiversity sensitive urban design. Conserv Lett. https://doi.org/10.1111/conl.12411

Geyle HM, Braby MF, Andren M, Beaver EP, Bell P et al (2021) Butterflies on the brink: identifying the Australian butterflies (Lepidoptera) most at risk of extinction. Austral Entomol 60:98–110

Gibson L, New TR (2007) Problems in studying populations of the golden sun moth, *Synemon plana* (Lepidoptera: Castniidae) in South-Eastern Australia. J Insect Conserv 11:309–313

Glatz RV, Leijs R, Hogendoorn K (2015) Biology, distribution and conservation of green carpenter bee (*Xylocopa aeratus*: Apidae) on Kangaroo Island, South Australia. Technical report, Department of Environment, South Australia

Grass I, Loos J, Baensch S, Batary P, Libran-Embid F et al (2019) Land-sharing/-sparing connectivity landscapes for ecosystem services and biodiversity conservation. People Nat 1:262–272

Haddaway NR, Brown C, Eggers S, Josefson J, Kronvang B, Randall N, Uusi-Kamppa J (2016) The multifunctional role of vegetated strips around and within agricultural fields. A systematic map protocol. Environ Evid 5:18. https://doi.org/10.1186/s13750-016-0067-6

Hall DM, Camilo GR, Tonietto RK, Olleton J, Ahrne K et al (2017) The city as a refuge for insect pollinators. Conserv Biol 31:24–29

Harvey JA, Heinen R, Armbrecht I, Basset Y, Baxter-Gilbert JH et al (2020) International scientists formulate a roadmap for insect conservation and recovery. Nat Ecol Evol 4:174–176

Haslett JR (1997) Suggested additions to the invertebrate species listed in Appendix II of the Bern Convention. Final report to the Council of Europe, Strasbourg

Haslett JR (2008) European strategy for the conservation of invertebrates. Nature and Environment, no 145. Council of Europe Publishing, Strasbourg

Haslett JR (2012) Development and future of conservation policy initiatives for insects and other invertebrates in Europe. In: New TR (ed) Insect conservation: past, present and prospects. Springer, Dordrecht, pp 317–337

Hill MJ, White JC, Biggs J, Briers RA, Gledhill D et al (2021) Local contributions to beta diversity in urban pond networks: implications for biodiversity conservation and management. Divers Distr 2021. https://doi.org/10.1111/ddi.13239

Hoffmann AA, White VL, Jasper M, Yagui H, Sinclair SJ, Kearney MR (2021) An endangered flightless grasshopper with strong genetic structure maintains population genetic structure despite extensive habitat loss. Ecol Evol 11:5364–5380. https://doi.org/10.1002/ece3.7428

Hogendoorn K, Glatz RV, Leijs R (2021) Conservation management of the green carpenter bee *Xylocopa aerata* (Hymenoptera: Apidae) through provision of artificial nesting substrate. Austral Entomol 60:82–88

Honan P (2008) Notes on the biology, captive management and conservation status of the Lord Howe Island stick insect (*Dryococelus australis*) (Phasmatodea). J Insect Conserv 12:399–413

Ikin K, Le Roux DS, Rayner L, Villasenor NR, Eyles K et al (2015) Key lessons for achieving biodiversity-sensitive cities and towns. Ecol Manag Restor 16:206–214

Kitching RL (1993) Rainforest canopy arthropods: problems for rapid biodiversity assessment. In: Rapid biodiversity assessment, proceedings of a workshop. Macquarie University, North Ryde, pp 26–30

Kitching RL (2004) Invertebrate conservation and the conservation of forests. In: Lunney D (ed) Conservation of Australia's forest fauna, 2nd edn. Royal Zoological Society of New South Wales, Mosman, pp 115–126

Kitching RL, Ashton LA (2013) Predictor sets and biodiversity assessment: the evolution and application of an idea. Pac Conserv Biol 19:418–426

Leandro C, Jay-Robert P, Vergnes A (2017) Bias and perspectives in insect conservation: a European scale analysis. Biol Conserv 215:213–224

Lewis OT, Basset Y (2007) Insect conservation in tropical forests. In: Stewart AJA, New TR, Lewis OT (eds) Insect conservation biology. CAB International, Wallingford, pp 34–56

Lewis J, Sands D (2021) *Anonychomyrma inclinata* sp. nov. (Hymenoptera: Formicidae): description, biology and interaction with the endangered bulloak jewel butterfly, *Hypochrysops piceatus* Kerr, Macqueen & Sands 1969 (Lepidoptera: Lycaenidae). Austral Entomol 60:461–472

Lundie-Jenkins G, Payne A (2000) Recovery plan for the Bulloak Jewel butterfly (*Hypochrysops piceata*) 1999-2003. Queensland Parks and Wildlife Service, Brisbane

Mata L, Garrard GE, Kutt AS, Wintle BC, Chee YE et al (2016) Eliciting and integrating expert knowledge to assess the viability of the critically endangered golden sun-moth *Synemon plana*. Austral Ecol 42:297–308

Milicic M, Popov S, Branco VV, Cardoso P (2021) Insect threats and conservation through the lens of global experts. Conserv Lett 2021:e12814. https://doi.org/10.1111/conl.12814

Mody K, Lerch D, Muller A-K, Simons NK, Bluthgen N, Harnisch M (2020) Flower power in the city: replacing roadside shrubs by wildflower meadows increases insect numbers and reduces maintenance costs. PLoS One 15(6):e0234327. https://doi.org/10.1371/journal.pone.0234327

New TR (1993) *Hemiphlebia mirabilis* Selys: recovery from habitat destruction at Wilsons Promontary, Victoria, Australia, and implications for conservation management. Odonatologica 22:495–502

New TR (2000) How to conserve the 'meek inheritors'. J Insect Conserv 4:151–152

New TR (2008) Legislative inconsistencies and species conservation status: understanding or confusion? The case of *Riekoperla darlingtoni* (Plecoptera) in Australia. J Insect Conserv 12: 1–2

New TR (2009) Insect species conservation. Cambridge University Press, Cambridge

New TR (2011) Butterfly conservation in South-Eastern Australia: progress and prospects. Springer, Dordrecht

New TR (2012) The Golden sun moth, *Synemon plana* Walker (Castniidae): continuing conservation ambiguity in Victoria. Vict Nat 129:109–113

New TR (2015) Insect conservation and urban environments. Springer, Cham

New TR (2021) Insect conservation and Australia's inland waters. Springer, Cham

New TR, Sands DPA (2004) Management of threatened insect species in Australia, with particular reference to butterflies. Aust J Entomol 43:258–270

New TR, Yen AL (2013) Invertebrate conservation in Australia: problems in policy and practice. Pac Conserv Biol 19:104–109

Priddel D, Carlile N, Humphrey M, Fellenberg S, Hiscox D (2003) Rediscovery of the 'extinct' Lord Howe Island stick insect (*Dryococelus australis* (Montrouzier) (Phasmatodea)) and recommendations for its conservation. Biodivers Conserv 12:1391–1403

Pywell RF, Heard MS, Woodcock BA, Hinsley S, Ridding L, Nowakowski NM, Bullock JM (2015) Wildlife-friendly farming increases crop yield: evidence for ecological intensification. Proc R Soc B 282:20151740

Richter A, Weinhild D, Robertson G, Young M, Edwards T, Hnatiuk S, Osborne W (2013a) More than an empty case: a non invasive technique for monitoring the Australian critically endangered golden sun moth, *Synemon plana* (Lepidoptera: Castniidae). J Insect Conserv 17:529–536

Richter A, Osborne W, Hnatiuk S, Rowell A (2013b) Moths in fragments: insights into the biology and ecology of the Australian endangered golden sun moth *Synemon plana* (Lepidoptera: Castniidae) in natural temperate and exotic grassland remnants. J Insect Conserv 17:1093–1104

Roitman M, Gardner MF, New TR, Nguyen JTT, Roycroft SJ et al (2017) Assessing the scope for genetic rescue of an endangered butterfly: the case of the Eltham copper. Insect Conserv Divers 10:399–414

Samways MJ (2020) Insect conservation. A global synthesis. CABI, Wallingford

Samways MJ, Simaika JP (2016) Manual of freshwater assessment for South Africa: Dragonfly Biotic Index. Suricata 2. South African National Biodiversity Institute, Pretoria

Sands DPA, New TR (2002) The action plan for Australian butterflies. Environment Australia, Canberra

Sands DPA, New TR (2003) Coordinated invertebrate surveys in Australia's National Parks: an important tool in refining invertebrate conservation management. Rec South Aust Mus Monogr Ser 7:203–207

Sands DPA, New TR (2013) Conservation of the Richmond birdwing butterfly in Australia. Springer, Dordrecht

Saunders ME, Barton PS, Bickerstaff JRM, Frost L, Latty T, Lessard BD, Lowe EC, Rodriguez J, White TE, Umbers KDL (2021) Limited understanding of bushfire impacts on Australian invertebrates. Insect Conserv Divers 14:285–293. https://doi.org/10/1111/icad.12493

Shaw MR, Hochberg ME (2001) The neglect of parasitic Hymenoptera in insect conservation strategies: the British fauna as a prime example. J Insect Conserv 5:253–263

Shrestha M, Garcia JE, Thomas F, Howard SR, Chua J et al (2021) Insects in the city: does remnant native habitat influence insect order distributions? Diversity 13:148. https://doi.org/10.3390/d13040148

Taylor GS, Braby MF, Moir ML, Harvey MS, Sands DPA et al (2018) Strategic national approach for improving the conservation management of insects and allied invertebrates in Australia. Austral Entomol 57:124–149

Theischinger G, Hawking J (2021) The complete field guide to dragonflies of Australia, 2nd edn. CSIRO Publishing, Clayton South

Trueman JWH, Hoye GA, Hawking JH, Watson JAL, New TR (1992) *Hemiphlebia mirabilis* Selys: new locations in Australia and perspectives on conservation (Zygoptera: Hemiphlebiidae). Odonatologica 21:367–374

Warren MS, Maes D, van Swaay CAM, Goffart P, Van Dyck H et al (2021) The decline of butterflies in Europe: problems, significance, and possible solutions. Proc Natl Acad Sci 118(2): e2002551117

White MJD (1956) Adaptive chromosomal polymorphism in an Australian grasshopper. Evolution 10:298–313

Wilkinson R (2014) Return of the phasmid. Media Dynamics, Australia

Yen AL, Butcher RJ (1997) An overview of the conservation of non-marine invertebrates in Australia. Environment Australia, Canberra

Zainol Abdullah WNZ, Apan AA, Maraseni TN, Le Broque AF (2014) Spectral discrimination of bulloak (*Allocasuarina luehmannii*) and associated woodland for habitat mapping of the endangered bulloak jewel butterfly (*Hypochrysops piceata*) in southern Queensland. J Appl Remote Sensing 8:083561

Appendix

This text has referred to the many threats to Australian insects, the shortcomings in documentation of native insects and in conservation awareness and progress, and the needs for urgent active steps to increase conservation effectiveness though both policy and practice. Many of those steps are principles familiar to all involved in conservation management in Australia, and the purpose of this Appendix is to summarise briefly some major themes and activities in conserving insects in their major habitats, with key references to further information and the development of the perspectives presented (for examples New 1995, 2009; Samways 2020). Most practical measures are inevitably general but collectively are the foundation for constructive progress and advance in transforming insect conservation from proposal to practice. They are the culmination of the complex process of conservation planning and transfer of ideas to operation. The 11 stages of conservation planning listed initially by Pressey and Bottrill (2008) and reiterated by Barmuta et al. (2011) (Table A.1), and acknowledgement that proactive measures are needed rather than simply 'locking up' areas to prevent their direct disturbance or exploitation, provide a constructive sequence and 'check list' as planning proceeds. Those activities widely involve detection, mitigation and prevention of threats, essentially safeguarding and/or restoring habitat quality.

The sequence implied may be varied or cycled as queries, new biological information, clarification of threats and conservation status, and assessment of outcomes necessitate revision and re-visiting earlier phases, and enhances the principles of wide consultation and collaboration, possible needs for compromise, and the critical ongoing dependence on sound information on which to base adaptive management. The unfamiliarity of insects and other invertebrates to many of those charged with conserving them, and the dearth of sources of advice, can limit development of conservation plans. Garnering all available advice on the biology of any focal insect species—likely to be very incompletely known - may be especially important in conservation planning. Wider surveys of insect diversity needed to characterise sites or systems or as a basis for monitoring may also need

© The Author(s), under exclusive license to Springer Nature Switzerland AG 2022

T. R. New, *Insect Diversity, Declines and Conservation in Australia*, Fascinating Life Sciences, https://doi.org/10.1007/978-3-030-90134-9

Table A.1 The 11 main stages of conservation planning, based on Pressey and Bottrill (2008) and Barmuta et al. (2011). The sequence of stages is not strictly linear, because some later phases may necessitate re-visiting earlier ones

Stage	Activities
Scoping and costing the planning process	Defining scope of planning, resources, staff and approach to each step of the process
Identifying and involving stakeholders	Involve, communicate with and build capacity of stakeholders who influence, affect or will be affected by implementation of the conservation management decisions
Identifying the context for conservation activities	Assessing the socioeconomic and political context, including constraints and opportunities for conservation management actions
Identifying conservation goals	Progressive refinement of values of stakeholders from a broad 'vision' to specific quantifiable goals
Collecting socioeconomic data	Collecting and evaluating spatially explicit data on tenure, extractive uses, costs, threats and existing management practices
Collecting data on biodiversity, other natural assets and ecosystem processes	Collecting and evaluating spatially explicit data on biodiversity pattern and process, ecosystem services and previous disturbances
Setting conservation targets	Translating goals into quantitative targets that reflect the conservation requirements for the targeted elements of biodiversity, natural assets and ecosystem processes
Assessing achievement of targets with existing management regimes	Assessing, by remote data and/or field survey, the achievement of targets under current management regimes
Modifying or selecting additional conservation management practices or procedures	Liaising with stakeholders to identify and prioritise the repertoire of management actions that achieve conservation targets while also accommodating other commitments, exclusions and preferences
Applying conservation actions to selected areas	Working through technical and institutional tasks involved in applying effective conservation actions to the areas identified in the planning process
Monitoring, evaluation and, where necessary, adaptation of conservation management actions	Close the adaptive management cycle by monitoring and evaluating agreed state and process variables to ensure the persistence of the identified conservation values

advice on the optimal methods to be used, and the difficulties and ambiguities they raise, as well as how the specimens are to be identified or sorted and, in due course archived for future use. In many situations, unexpectedly high costs of undertaking and analysing insect surveys may limit the exercise, but any approximations (such as by refuge in higher taxa separations alone) may need very careful evaluation. Referring to freshwater ecosystems, but applicable equally to terrestrial environments, 'Systematic assessments of ... biodiversity (which, by definition, include the less charismatic taxa) will be an extra layer of complexity that some managers and stakeholders would like to avoid' (Barmuta et al. 2011). Progressive familiarity with insects can lessen that reluctance, but the complexity is likely to persist at least until practical experience has been obtained.

Considerations for Insect Species Conservation

Species-level conservation is founded in understanding the focal species, its needs, and what is needed to sustain or restore it (New 2009). If the species is 'listed' with a defined conservation status, the nomination documents together with any action plan or other management pro forma arising from the listing action are likely to be the most useful 'starting points' in information gathering, with additional effort to trace more recent developments that, in some cases, lead to modifications of the initial perspective. The breadth of information needed may be satisfied in some cases largely from published papers and internal reports of conservation agencies, but significant gaps may remain, and necessitate field investigations.

Especially when such regulatory background is unavailable, such as for unlisted species, a checklist of points needed to justify conservation action—modified and augmented as necessary—might include: Species' identity (is it unambiguous?; may need good photographs or voucher specimens [if necessary taken under permit] for validation by specialist or at State Museum); recognition (can it be recognised easily by non-specialists, including citizen scientists, or is it likely to be confused with other taxa?); has it wider interest or significance (such as being locally endemic, taxonomically isolated or relict, having cultural value, or other); conservation status (has it been allocated an IUCN Red List category, or any other reliable conservation designation and, if so, is the evidence available for review?); historical perspective (summarise historical information on distribution and conservation interest: are there suggestions/evidence of earlier declines or losses and if so, can these be attributed to any cause?; has there been earlier conservation attention?); amenability to detection and survey (is it diurnal/nocturnal and sufficiently conspicuous for survey by relatively simple methods such as visual surveys/searches, by non-specialists, or are more complex measures needed to detect and enumerate it?); knowledge of early stages (are adults and larvae clearly associated, and larvae recognisable?); knowledge of life history (is pattern of seasonal development and appearance [such as flight season] defined and predictable); population structure (is the population closed, open, or a metapopulation?); resources (are critical resources such as particular food plants or prey known, and can they be surveyed?; are any particular utilities, such as aestivation, nesting or patrolling sites, needed?); threats (current or projected) (can they be defined and impacts assessed; is prevention or mitigation possible and, if so, how?; are they site-restricted or more widespread?); arena of activity (is conservation site-restricted, if so is/are the site(s) secure from long-term conservation activity? Will conservation be wholly in situ, or are captive breeding, translocations and related activities likely to be needed?); specific management measures (define, integrate and cost); support (is the focal species politically or publicly 'controversial', or is conservation need accepted widely?; does community interest indicate likely citizen science interest and help?; can the conservation plan be supported logistically, including expertise and finance?; are further regulatory measures needed, such as nomination for listing?); related or co-occurring taxa (are any close relatives of conservation interest?; do other insects [or other taxa] of conservation concern occur on the same sites or use the same resources?: if either of these, are there opportunities for joint conservation planning and management?).

Considerations for Wider Insect Conservation

Changing emphasis from the fine-filter approach to meso- or coarse-filter insect conservation embraces needs for wider insect biodiversity, with primary interests on the characteristic environments or habitats needed, at scales from individual sites, particular topographical or ecological features or communities to landscape considerations. Individual notable species may contribute to the wider values of those entities, as enhancing conservation significance. Several examples of such flagships are discussed in Chap. 10, and threatened insects may combine with other taxa as joint ambassadors for key environments. Some native grasslands in south east Australia have been disparagingly noted as 'harbours of deformity', reflecting their importance in co-hosting a 'mouthless moth' (*Synemon plana*, p. 200) and two threatened reptiles known as the 'earless dragon' and 'legless lizard', respectively, and also restricted to those environments.

The following exemplify themes relevant in conserving insect habitats for biodiversity or more individual focus, with the goal of preserving, sustaining, restoring and conserving at scales from local to landscape and—if necessary, creating new habitats. Emphasis is thus predominantly on the hosting environments and assuring resource needs for insects rather than on the insects themselves. For such habitat focus, pointers include: definition (can the area/ecosystem/site of interest be defined and delimited unambiguously? Is it a single unit or are multiple 'patches' [such as for a metapopulation] involved?); tenure(s) (is ownership/proprietorship defined unambiguously and clearly, and assured for longer-term conservation management? If not, is achieving this a practical proposition? Are all stakeholder groups involved actively in conservation activity?); level of protection (is the area formally protected, under covenant or restriction of any kind, with private or public access, targeted for change, and with single or multiple current uses?); interests (are there competing or conflicting interests in the area, such as for development, exploitation, reservation or involving any threats likely to change the area and which may need to be accommodated or countered in pursuing conservation priority); significance (why is the unit important?: remnant, long-isolated, 'one-of-a-kind', unique associations of vegetation or insect fauna [permanent or seasonal], unusually high diversity, historically important study site, type locality, cultural value, or other factor(s) implying need for conservation); rationale (summarise intrinsic and extrinsic values; specify conservation goals and how these may be pursued; are there clear priority needs or actions?; is management responsibility clearly allocated, for example to a particular agency or body?); pressures/threats from enveloping environment (assess impacts from adjacent agricultural, industrial, urban surrounds; edge effects for linear sites such as roadsides and riparian strips); size and position (is area sufficiently large to conserve representative insect diversity—is there an ample 'core area' free from edge effects?; can it be buffered against intrusion?; is it likely to be functionally isolated [such as a single urban remnant of natural vegetation, or a pond] or adequately connected to other similar ecosystems—if not, can connectivity be enhanced?).

Considerations for Habitat Management and Threat Mitigation

Whether for single insect species focus, or for wider conservation perspectives, habitat management is the prevalent theme in insect conservation, and the goals of preserving, sustaining and restoring habitat link closely with removal, prevention or mitigation of threats to those environments. Anthropogenic impacts are diverse, and are superimposed on more natural processes such as vegetation succession that may markedly alter the suitability and carrying capacity of an area—and need to be addressed in management for species affected adversely. The very diversity of insect habitats in Australia, spanning aquatic (lotic, lentic, intermediate) and terrestrial (desert to forest regimes) across latitudinal and elevational gradients, and natural to highly anthropogenic environments is daunting when seeking management generalisations, and proper treatments for each such biome have been explored extensively. Nevertheless, some general principles emerge, based mainly in those of wider landscape ecology, and often predicated on suitable habitat patches being distributed unevenly across an enveloping contrasting matrix that is far less habitable and may impede movement of insects.

Consideration of multiple sites usually involves dealing with multiple tenures and management responsibilities/inputs, priorities and levels of security, as well as a range of sizes, conditions and threats, so that deliberate incorporation of all available patches into a conservation plan is rarely possible or, even, desirable. Wide consultation is almost always needed to explore the range of possibilities for modifications, management, and/or conflicting interests that must be resolved or accommodated, as well as to explore what is needed for each candidate site in terms of threat reduction, remediation, additional protection and continuing management. Goals may involve selecting a series of focal sites within the extensive landscape matrix, on features such as size (larger sites, buffered and with secure core areas largely free from edge effects), shape (linear sites less preferable than circular sites, for edge effects, as above), ecological uniqueness in beta diversity (p. 55), high species richness, presence of designated threatened species, relative freedom from alien invasive weeds and other species, naturalness and low threat levels and potential increase of threats, feasibility of management, and connectivity. A preliminary series of sites (with potential to add others) can form the basis of a planning ecological network, emphasising retention and restoration of heterogeneity and promotion of connectivity by reducing effective isolation of each. Construction of ‘stepping stones’ or seeking these as additional protected areas between present patches can also be explored.

References

Barmuta LA, Linke S, Turak E (2011) Bridging the gap between ‘planning’ and ‘doing’ for biodiversity conservation in freshwaters. Freshw Biol 56:180–195
New TR (1995) Introduction to invertebrate conservation biology. Oxford University Press, Oxford
New TR (2009) Insect species conservation. Cambridge University Press, Cambridge
Pressey RL, Bottrill MC (2008) Opportunism, threats, and the evolution of systematic conservation planning. Conserv Biol 22:1340–1345
Samways MJ (2020) Insect conservation. A global synthesis. CABI, Wallingford

Index

© The Author(s), under exclusive license to Springer Nature Switzerland AG 2022

T. R. New, *Insect Diversity, Declines and Conservation in Australia*, Fascinating Life Sciences, https://doi.org/10.1007/978-3-030-90134-9